Reliability in Cognitive Neuroscience: A Meta-Meta-Analysis

Books Written by William R. Uttal

Real Time Computers: Techniques and Applications in the Psychological Sciences

Generative Computer Assisted Instruction (with Miriam Rogers, Ramelle Hieronymus, and Timothy Pasich)

Sensory Coding: Selected Readings (Editor)

The Psychobiology of Sensory Coding

Cellular Neurophysiology and Integration: An Interpretive Introduction.

An Autocorrelation Theory of Form Detection

The Psychobiology of Mind

A Taxonomy of Visual Processes

Visual Form Detection in 3-Dimensional Space

Foundations of Psychobiology (with Daniel N. Robinson)

The Detection of Nonplanar Surfaces in Visual Space

The Perception of Dotted Forms

On Seeing Forms

The Swimmer: An Integrated Computational Model of a Perceptual-Motor System (with Gary Bradshaw, Sriram Dayanand, Robb Lovell, Thomas Shepherd, Ramakrishna Kakarala, Kurt Skifsted, and Greg Tupper)

Toward A New Behaviorism: The Case against Perceptual Reductionism

Computational Modeling of Vision: The Role of Combination (with Ramakrishna Kakarala, Sriram Dayanand, Thomas Shepherd, Jaggi Kalki, Charles Lunskis Jr., and Ning Liu)

The War between Mentalism and Behaviorism: On the Accessibility of Mental Processes

The New Phrenology: On the Localization of Cognitive Processes in the Brain

A Behaviorist Looks at Form Recognition

Psychomyths: Sources of Artifacts and Misrepresentations in Scientific Cognitive neuroscience

Dualism: The Original Sin of Cognitivism

Neural Theories of Mind: Why the Mind-Brain Problem May Never Be Solved

Human Factors in the Courtroom: Mythology versus Science

The Immeasurable Mind: The Real Science of Psychology

Time, Space, and Number in Physics and Psychology

Distributed Neural Systems: Beyond the New Phrenology

Neuroscience in the Courtroom: What Every Lawyer Should Know about the Mind and the Brain

Mind and Brain: A Critical Appraisal of Cognitive Neuroscience

Reliability in Cognitive Neuroscience: A Meta-Meta-Analysis

Reliability in Cognitive Neuroscience: A Meta-Meta-Analysis

William R. Uttal

The MIT Press
Cambridge, Massachusetts
London, England

© 2013 Massachusetts Institute of Technology

All rights reserved. No part of this book may be reproduced in any form by any electronic or mechanical means (including photocopying, recording, or information storage and retrieval) without permission in writing from the publisher.

MIT Press books may be purchased at special quantity discounts for business or sales promotional use. For information, please email special_sales@mitpress.mit.edu or write to Special Sales Department, The MIT Press, 55 Hayward Street, Cambridge, MA 02142.

This book was set in Stone Sans and Stone Serif by Toppan Best-set Premedia Limited, Hong Kong. Printed and bound in the United States of America.

Library of Congress Cataloging-in-Publication Data

Uttal, William R.
Reliability in cognitive neuroscience : a meta-meta analysis / William R. Uttal.
  p. ; cm.
Includes bibliographical references and indexes.
ISBN 978-0-262-01852-4 (hardcover : alk. paper)
I. Title
[DNLM: 1. Mental Processes—physiology—Review. 3. Brain—physiology—Review. 3. Brain Mapping—Review. 4. Cognition—Review. 5. Meta-Analysis as Topic—Review. 6. Reproducibility of Results—Review. WL 337]

612.8'233—dc23
2012016250

10 9 8 7 6 5 4 3 2 1

For Mitchan

We face an abundance of information. Our problem is to find the knowledge in the information. We need methods for the orderly summarization of studies so that knowledge can be extracted from the myriad individual researches.
—G. V. Glass (1976, p. 4)

Mental faculties are notions used to designate extraordinarily involved complexes of elementary functions. . . . One cannot think of their taking place in any other way than through an infinitely complex and involved interaction and cooperation of numerous elementary activities, with the simultaneous functioning of just as many cortical zones, and probably of the whole cortex, and perhaps also including even subcortical centers. Thus, we are dealing with a physiological process extending widely over the whole cortical surface and not a localized function within a specific region. We must therefore reject as a quite impossible psychological concept the idea that an intellectual faculty or a mental event or a spatial or temporal quality or any other complex, higher psychic function should be represented in a single circumscribed cortical zone, whether one calls this an "association centre" or "thought organ" or anything else.
—K. Brodmann (1909; translated and edited by Garey, 1994, p. 255, as located by Ross, 2010)

It's easy to sell simple stories with memorable take-home conclusions (the amygdala is the seat of fear, etc.), but it's harder for people to understand and accept more complex models, I think.
—Tor Wager (personal communication, 2011)

Contents

Preface   ix
Acknowledgments   xv

1 Meta-analysis: The Idea   1

2 Meta-analysis: The Methodology   27

3 On the Reliability of Cognitive Neuroscience Data: An Empirical Inquiry   79

4 Macroscopic Theories of the Mind-Brain   139

5 Current Status and Future Needs   185

Notes   197
Bibliography   209
Name Index   223
Subject Index   229

Preface

This book is a study of reliability, of replicability, and of consistency. As such, it is a test of the merit and validity of a body of scientific research that has been playing an increasingly central role in the field of cognitive neuroscience. The general question is: Do macroscopic brain images produced by a rapidly evolving technology reflect, correlate, or represent cognitive processes? The specific question asked in this book is: Can this question be better answered by pooling the data from a number of experiments than by depending on the results from a single one? The main thesis presented in this book is that despite its unchallengeable utility in anatomic and physiological applications, brain imaging research has not provided consistent answers to the general question for cognition. Furthermore, the answer to the specific question is that little is gained by pooling brain image data. These theses are based on a review of the empirical literature carried out at several levels of comparison. It is shown that there is great variability not only among individual subjects and experiments but also in the most recent approach to pooling data—meta-analysis.

Because of the importance of reliability in scientific research, any lack of consistency that is exhibited by the empirical evidence in this field has profound implications. This inconsistency suggests that cognitive neuroscientists using this approach have not yet proven their interpretations and theories of the relations between this kind of brain activity and cognitive processes. What may have appeared to be correlations and relations might have only been illusions of association stimulated by a host of attractive and seductive features of brain imaging technology. Underlying these "illusions" may be a proclivity on the part of cognitive neuroscientists to see order in what are essentially random responses and to have mistakenly accepted marginal statistical criteria as proofs. Although there have been indications in the past that all was not well in correlative studies of brain image and cognitive processes, the evidence for inconsistency and unreliability is increasing year by year.

The reasons that associating brain images and cognitive processes have become such a popular methodology in cognitive neuroscience are both understandable and opaque. Most important of all of the understandable reasons is that the mind-brain relationship is the preeminent question of human existence, and the urge to solve it is intense. Of equally consequential import is the fact that we still have no idea how the wonders of consciousness emerge from the electrochemical interactions among the innumerable neurons of the brain. Therefore, we reach out to anything that has even a remote chance of explaining this mysterious and wondrous process. Throughout their history, psychologists and neuroscientists have sought any new technology in the hope that we would be able to find some entrée into understanding how the brain and the mind are related. Of the basic premise that mind is a brain process, there is no controversy in scientific circles; how and at what level of analysis it happens, there are barriers to understanding galore.

It is, for example, possible that we are asking the wrong questions. By attaching our research to macroscopic brain imaging procedures, we may be operating at the wrong level of analysis. Indeed, it is possible that the "where" question (i.e., in what parts of the brain do activations correlate with cognitive processes?) may be what philosophers would call a bad question. This would certainly be the case if mind emerged from the actions at the microscopic neuronal level—a highly plausible and likely possibility, but one that is still unproven. No matter how plausible the alternatives may be, there has never been an empirical test of this hypothesis that meets strict criteria of proof.

Another reason that the field has prospered in the face of inconsistent and contradictory data is that the complexities of the brain and the uncontrolled nature of the cognitive processes generated by specific experimental conditions will always permit some sort of a differential response no matter how few or many the subjects or how many experiments are meta-analyzed. Indeed, it is extremely difficult to carry out a brain imaging experiment that results in no response difference between control and experimental conditions.

In our zest to understand, other factors have influenced the reification of macroscopic brain activity and led us to see order where there may actually only be randomness. Complexity creates opportunities to misperceive reality; attractive pictures exert a compelling argumentative force when they serve as adjuncts to the relative barrenness of tables and charts. Our compulsive need to find order where there may be none is a human trait that can go astray. Of lesser importance is that an enterprise like this can feed on itself, becoming a venture with its own culture, its own vested interests, and a continuing need to support an ever-increasing investment in an expensive and growing material infrastructure that must be both financed and fed by public relations hyperbole. The end result is a misdirection of resources and students from what might actually be a more logical and relevant approach to human behavior.

This book is intended to be a further step in my ongoing effort to evaluate the validity of relations drawn between brain images and cognitive processes. In earlier books (Uttal, 2001, 2009, 2011), I considered the question from three different points of view. In the first of these three books, I argued on essentially lexicographic and technical grounds that the brain imaging research of the previous decade did not support the narrow localization of cognitive processes in particular regions of the brain. This suggestion, made by others as well as by me, has now been widely accepted by cognitive neuroscientists.

In the second book, I considered the problem of distributed brain responses in greater detail. During the first decade of the twenty-first century a massive amount of research made it clear that responses of the brain to almost any cognitive process were widely distributed around many if not most parts of the brain. Furthermore, research during the last two decades has shown that there was considerable inconsistency among the reports of the distributed brain locations. That is, although most current researchers now agree that it is a broadly distributed pattern of brain responses that answers the "where" question, there was and is considerable disagreement concerning the nature of that distribution. This is a major controversy currently brewing in cognitive neuroscience.

In the third book, I concentrated on interexperiment variability. An extensive review of the empirical literature showed that cognitive neuroscience has a long history of inconsistent results and abandoned, rather than solved, problems. When scientists are faced with situations like this in which data are very variable and the signal-to-noise levels are poor, the approach is to apply statistical methods. Indeed, this is what had happened in the first decade of the twenty-first century to the field of brain imaging. There has been a rapid expansion in what are called meta-analytic techniques. Just as statistical methods pool data obtained from a sample of subjects within the confines of a single experiment, meta-analyses seek to pool or combine the findings from a number of experiments.

The idea seems straightforward; however, as we see in this current book, the task of pooling data from different experiments is extremely complex, difficult, and challenging. As a result, consistency among comparable meta-analyses is rarely obtained. There remains considerably uncertainty whether or not even these massive pools of data produce consistent and psychobiologically plausible answers to the question of the relations between brain mechanisms and cognitive processes.

Consistency is primarily an empirical question. We may well find in the future that certain patterns of distributed activation are actually reliable correlates of cognitive processes. However, there is also the possibility that the outcomes of meta-analyses may continue to be so irregular that the whole question of spatial representation of cognitive processes may have to be rethought.

Furthermore, even if we achieve some kind of replication, answers to the question of what these patterns mean remain speculative. For example, it is possible that cognitive processes are encoded by the universally distributed actions and interactions of the great neuronal networks of the brain. Such a hypothesis is not necessarily contradictory to the distributed localization hypothesis; it may simply be a matter of different levels of inquiry. On the other hand, it may be that the questions asked at the macroscopic level probed with fMRI techniques are bad questions because this is not the meaningful level of analysis at which mind is instantiated. The technical question then arises—what, then, do the correlated effects of changing blood distributions mean in the context of the great question: How does the brain make the mind?

It is not likely that any of these grand questions will be answered by even the most advanced meta-analytic technique. Nevertheless, one of the key issues in carrying out this new review is whether or not, these advanced techniques are leading us toward consistent answers to even the most basic empirical conundrums. In this book, I examine and compare results at several levels of analysis culminating in meta-analyses to see if higher-level data pooling produces more reliable and consistent answers to the role that various regions of interest play in mind-brain relationships. In short, my ultimate goal is to carry out a meta-meta-analysis. If it turns out that consistency cannot be found at this highest level of data pooling, then there is an urgent need for us to reevaluate the entire enterprise.

My approach is going to be to raise questions and then to see what in the empirical literature pertains to them. In doing so, the selection of reports is going to have to be incomplete. I apologize in advance to the authors of those articles that I miss. It was not intentional; no one could possibly review the entire corpus of publications in this field fully. I am sure there are some very relevant studies that might well have stood in place of others that I did find. Others might have reached different conclusions. Whether or not they might have changed my mind about the state of this field is problematic.

There is another issue that I should lay out at the beginning of this book. Often I found myself in disagreement with the interpretation of the data in a particular article. Notwithstanding the original author's conclusions and theories, in a number of such cases I have taken the liberty of providing my own interpretation of what the offered results mean. This is a necessary strategy in a world of vague stimuli and variable responses.

Because of the current state of brain imaging for cognitive neuroscience, this book has to be a collection of questions rather than a collection of answers. I hope that my emphasis is on the unknowns as I take a few modest steps toward answering some of them.

Acknowledgments

Writing a book like this one is essentially a solitary endeavor. However, it is done in a context of work by a large number of scientists who provide the raw material for my discussions. I acknowledge my debt to all of them, and, even when disagreeing, I hope that I never the crossed the boundary between substantive controversy and ad hominem remarks. This is a field of great passion and vested interest, and it is all too easy to demonize an intellectual opponent. If I have come near that boundary, I apologize in advance.

There are some more immediate contributors whom I also want to acknowledge. Jeff Cochran at the U.S. Air Force Institute of Technology at Wright-Patterson Air Force Base in Ohio and John Reich of Arizona State University provided some useful advice on statistical matters. I am also grateful to two anonymous reviewers and to Jeremy Skipper of Hamilton College for their useful and constructive critiques of a draft of this book. Guiding the editorial process has been Philip Laughlin of MIT Press whose cordial and supportive role cannot be minimized. Phil and his staff did a wonderful job producing this book as they did on my previous work, *Mind and Brain: A Critical Appraisal of Cognitive Neuroscience*.

I am grateful to Arizona State University's College of Engineering for making services and space available to me long after my retirement. I am also still enjoying the hospitality and intellectual stimulus of the University of Hawaii during the last 10 summers. Many thanks to Professor Pat Couvillon for her administrative and organizational skills in making these visits possible.

Finally, as ever, it is my dear wife of all these years—Mitchan—who has provided the loving and supportive environment that has made everything I have done possible.

1 Meta-analysis: The Idea

1.1 Introduction

This book concerns the meta-analyses of brain imaging data—a currently popular approach to study the relation between brain responses and cognitive processes. It is a part of the continuing search for an answer to what many believe is the most important and complex scientific question of all time—how does the brain make the mind? In the last two decades, the effort to answer this question has accelerated enormously with the introduction of new measurement systems for examining what are thought by many to be the neural correlates of cognitive activity. The allocation of scientific resources to this quest has been extraordinary in historical terms. The number of scientific reports that document one or another aspect of the question seemingly grows exponentially every year. An enormous database is emerging from this effort. Probably over a thousand papers are currently being published every year using brain imaging to study the problem.

However, if one were to pick out a single word to characterize the current state of cognitive neuroscience data obtained with brain imagining systems, it would probably have to be "inconsistent." Although there have been a plethora of experimental reports purporting to show activations in particular parts of the brain when particular cognitive processes are executed, only a few consistently show reliable results that remain stable from one experiment to another or even from one subject to another. The reasons for this extreme variability are beginning to become clear, but the problem of what to do about them has created a major preoccupation for modern cognitive neuroscience. The hope was that, as in the other sciences, we would be able to find some way to merge the results of a number of disparate experiments to produce a better answer to the question of mind-brain relations than can be drawn from a single experiment. Given the relatively large differences that even comparable experiments are producing, this has become a very challenging task. This is particularly so for experiments using systems that are mainly able to answer only the "where" question—a limitation of most brain imaging techniques. That is, how do you combine data that are primarily presented in the form of spatial locations?[1]

Our ability to combine or pool spatial data from brain images is extremely important for two reasons. First, it is necessary to seek out common outcomes in a context of significant variability. This is the normal progress of science; typically, as more and more data are collected, a more precise estimate of the "central tendency" of a sample of data is obtained. From the aggregation of increased amounts of data, it is hoped that general laws and principles will emerge.

Second, it is important to be able to distinguish between the spurious and the real in the present context. One problem is that data obtained in a group of experiments are often extremely inconsistent, diverging from rather than converging toward a stable general law or interpretation. Such a result at least raises the possibility that there is no "central tendency" or singular answer to questions concerning the role of particular parts of the brain in a given cognitive task. It is typical (as we see in later chapters) that as more data are collected, ever greater variability of the neural responses is observed.

There is a real possibility, therefore, that we are ascribing much too much meaning to what are possibly random, quasi-random, or irrelevant response patterns. That is, given the many factors that can influence a brain image, it may be that the cognitive states and the brain image activations are, in actuality, only weakly associated. Other cryptic, uncontrolled intervening factors may account for much, if not all, of the observed findings. Furthermore, differences in the localization patterns observed from one experiment to the next nowadays seem to reflect the inescapable fact that most of the brain is involved in virtually any cognitive process. The idea that there are specialized and circumscribed brain regions associated with particular cognitive processes has dominated cognitive neuroscience as far back as the Middle Ages (see Kemp, 1996), the Renaissance, and well into modern times as epitomized by the failed idea of phrenology posed by Franz Joseph Gall (1758–1828) and Johann Spurzheim (1776–1832).

This basic postulate of localized brain regions for specific cognitive processes, however, seems less and less tenable as the years go by. Whatever differences exist between experiments may be due to slight differences in controlled and uncontrolled variables rather than to any compartmentalization of either cognition or brain activity. Although none of us want to accept the remote possibility that our findings are really examples of reading order into random disorder, given the high degree of variability of brain activation patterns and the difficulty of controlling all of the influential variables, such a possibility cannot be completely rejected.

Indeed, relevant data pooling may turn out to be better suited to rejecting questionable localization theories than for converging on models in which particular brain regions are associated with particular cognitive functions. There is an enormous burden on scientists to determine if a surprising and unexpected result is a real phenomenon or whether it represents nothing more than an artifact resulting from inadequately designed or underpowered experiments; attempts to combine (i.e., to meta-analyze) data might, it has been suggested, be the way to accomplish that task.

The special hope is that exploiting the methods and outcomes of data pooling can also eventually allow progress to be made in understanding how the activity of the brain corresponds (or, in some cases, does not correspond) to psychological processes. This book is aimed at exploring the findings, methods, and means by which the variable and abundant findings of brain imaging experiments can be used to determine if this hope can be fulfilled.

However, there are many caveats in pursuing this line of inquiry. It is not even certain at this stage of the game that we are going in the right direction. It may be that we are looking at the wrong level of analysis with macroscopic brain imaging techniques such as the functional magnetic resonance imaging (fMRI) technology. It is not at all clear, for example, that psychological processes map directly onto places in the brain in the simple way implicitly assumed by many cognitive neuroscientists these days. The still influential foundation postulates of (1) modular psychological functions and (2) localized brain function may have forced us into a kind of experimental paradigm in which it is assumed that particular mental processes are encoded or represented by macroscopic brain locations or mechanisms. To pursue this quest in this uncritical and possibly incorrect direction, however, may detour us from studying the more microscopic level of analysis that seems more likely to be the seat of mind-brain equivalences.

It seems increasingly likely, therefore, that the two assumptions previously governing cognitive neuroscience research (mind is modular, and these functional modules can be localized on restricted locations of the brain) may both be wrong.[2] Modern psychological research is increasingly showing that such cognitive processes as emotion, perception, learning, and attention are far more tightly interwoven with each other than it may have seemed to some of our predecessors. The Cartesian idea that we can, indeed must, study learning or emotion independent of each other (and of most other cognitive processes) is now being increasingly challenged. The great conundrum, as one recent scholar[3] suggested, is this: "Are we cutting the mind into pieces at the real 'joints' or making arbitrary cleavage points because of the clumsiness of our research tools, especially our cognitive taxonomies?"

Similarly, modern neuroscience research is increasingly showing that activation areas on the brain associated with a cognitive process are far more widely distributed than had been thought only a decade or so ago. Indeed, it now seems very likely that most of the brain is active in almost any cognitive process. If both of the currently prevailing assumptions (cognitive modularity and localized brain regions) are shown to be incorrect, much of the brain imaging research of the last two decades may turn out to have been misdirected in principle as well as confused by the conflicting and inconsistent empirical findings.

Currently, the tide is turning away from simplistic ideas of narrow regional localization to more complex idea of contingencies among a number of brain regions, that is, of coordinated and interconnected systems. (Psychologists may also be moving away from cognitive modularization, but this is less clear; the concept of mental "faculties" or "modules" still exerts a powerful influence on psychological thinking.) Thus, we now tend to ask which of a number of brain regions seem to be operating Thus, we tend to ask: Which of a number of brain regions seem to be operating together during a given cognitive process, and how are they interconnected? The key question has become "How is 'a coordinated system of brain loci' associated with a cognitive task?" as opposed to asking "Which of a few narrowly circumscribed locations of the brain represent that process?"

As in any other scientific enterprise, the best way to determine if the current paradigm of modularization and localization is, in fact, misdirected is to look at the empirical outcomes. What constitutes an "empirical fact," however, in this field is not always clear given the variability of the resulting observations. Unfortunately, many experiments that have been reported in this burgeoning literature have been done with designs that are of insufficient statistical power[4] to provide the basis for confidently establishing a "fact." Furthermore, brain imaging is an expensive and time-consuming approach to the study of the mind-brain problem. As a result, sample size is often constrained by practical economic considerations, and, therefore, the statistical power of many experiments may be unacceptably low. In many articles, the reported findings are based on a modest number of subjects, all too often fewer than a dozen. In a domain in which there is such variability among individual subjects and so many uncontrolled variables that may affect the determination of the activation areas, such small numbers may lead to gross misestimates of the actual relations between cognitive and neurophysiological states.

Therein lies the possible advantage of meta-analytic methods that combine, pool, or average data from a number of comparable experiments. The hope is that sufficiently "powerful" protocols, not always to be found within the confines of a single experiment, can be generated by pooling experimental results in much the same way we can add subjects in a conventional psychological experiment to enhance our confidence in the forthcoming findings. Meta-analyses of this kind have become especially important in cognitive neuroscience research in recent years due to a proliferation of experimental reports, many of which are underpowered, leading to inconsistent results.[5]

Data pooling as a means of effectively increasing the power of studies is the basic statistical philosophy for improving the precision of estimates in many kinds of research. The principle is simple: data are pooled from a number of different sources or observations in order to improve confidence in the most likely value of the parameter under investigation. This same fundamental philosophy is embodied in the idea of a meta-analysis. Rather than simply pooling the outcomes of individual observations within an experiment, the already integrated results of a number of experiments are pooled or combined to produce a better estimate of the answer to the question being asked or to evaluate some hypothesis that may well have been supported by a least some of the individual experiments.[6,7]

In principle, the meta-analytic approach to studying the neural loci of cognitive processes should work; however, there are many practical problems with it, some of which are unique to its application to brain imaging studies. In the next chapter, I detail some of these problems and the sources of bias that produce them. However, there is a very serious empirical problem with neuroscientific meta-analyses that must be emphasized at the beginning of this discussion—pooling the results of several experiments often leads to divergences or increases in the variability of brain areas rather than a convergence onto a few localized regions. Specifically, meta-analyses of brain images increasingly often lead to more widely distributed activations than is suggested by any of the individual experiments. Indeed, as I tried to establish in my earlier work (Uttal, 2011), the most compelling conclusion from meta-analyses is that the more we pool or combine data, the more distributed are the regions of the brain that seem to be involved in even the simplest cognitive process—if there is any such thing as a simple cognitive process.

Furthermore, there are large individual differences observed in this research program that must be taken into account. As we see in chapter 3, it is becoming increasingly clear that variability actually increases as data are sequentially pooled. That is, individual subjects exhibit some variability from day to day, intersubject variability is even greater, and there is ever-increasing variability as more and more data from groups of experiments are pooled into meta-analyses.[8]

This general result—increasing distribution of activated brain regions with increased sample size—appears to be typical for any experiment that seeks to answer the question: What region or regions of the brain are involved in the representation or encoding of a particular cognitive process? Consequently, it raises serious questions about the basic assumption that we will progressively converge on a more precise answer by sequentially pooling or meta-analyzing data. Pooled data from a meta-analysis may, therefore, also imply that the "where" question itself may be a "bad question."[9] The brain, our meta-analyzed data are forcefully suggesting, is functioning in a much more holistic fashion than the conventional assumptions of localized brain functions and modular cognitive modularity would have had us believe until recently.

There are several ways in which this increasing divergence of brain imaging data, rather than convergence with progressive data pooling, can be explained. The first is the conventional one; namely, for a variety of reasons, individual experiments produce valid but variable responses. That is, the localized activation regions stimulated by any given cognitive task are, in psychobiological fact, driven by the nature of the task. According to this view, the responding brain regions truly encode, correspond to, or instantiate the cognitive process. Variability and divergent distribution, in this context, arise because the response of the brain is exquisitely sensitive to the parameters of the experimental protocol. Thus, there is a somewhat different pattern of localized brain responses as each additional experiment is carried out. Following this interpretation, as we simply accumulate the results of a number of experiments, more and more brain regions are activated, and the relevant neural system seems to involve ever more widely distributed regions. In psychobiological fact, there *is* some kind of specialized functional localization of cognitive processes. However, it is hidden because supposedly comparable experiments are actually not so comparable; instead, the observed differences are mainly due to our limited control of critical experimental conditions. Variability, from this perspective, therefore, is an artifact of the imprecision of our experimental protocols, poor stimulus controls, and the brain's ability to adapt to that lack of control. In sum, the implication is that the localization postulate is actually a valid description of brain organization, but it is often obscured by the lack of control over stimulus and task conditions. From this perspective, meta-analyses of brain images may be intrinsically more noisy than other kinds of data and, thus, actually be less reliable than meta-analyses of other kinds of scientific problems.

A second kind of interpretation of the apparent divergence or broadening of the distribution of brain responses as data are pooled is that broad rather than localized distribution is also, in empirical fact, the actual means by which the brain encodes or represents cognitive processes. From this perspective, localized responses are artifacts resulting primarily from the high cutoff criteria applied by investigators. High thresholds artificially emphasize a few large peaks of activation and ignore lower-level activity. This means that what are usually reported as activation "peaks" are actually only the maximum values of much larger areas or regions of activation. Thus, a localized peak is a false characterization of a much broader region of activity; this explanation suggests that the whole idea of localization is misleading. The implication in this case is that individual experiments that falsely indicate localized responses are obscuring the reality of distribution by arbitrary setting of the thresholds.

The third possibility is one that most of us would like to sweep under the table; namely, that the whole idea of brain image correlates of cognitive processes is an artifact—an illusion! What may appear to be significant results are in large part random and quasi-random responses that are more the result of the neuroanatomical organization of the brain than of the details of the cognitive activation. In other words, our interpretations of our data inject illusory order into what is actually an unstructured response. Our task is not, in this case, to explain why there is such variability of "truly" localized activations (explanation number 1) or why the true global distribution of brain activations has been obscured in individual experiments (explanation number 2); instead, it is to explain why so many experimenters report some kind of orderly responses when, in fact, there is no order.

There are many complex issues to be pondered in this context. Is it possible that "cognitively" driven brain images are actually the result of some cryptic processes that have not yet been identified and actually have little to do with cognition per se? Alternatively, is it possible that the variation among individuals is so great that it will not be possible to answer the specific question of representation for any particular cognitive process from one individual to the next? Are our methods for pooling data and small sample sizes introducing an appearance of orderliness that disappears with larger sample sizes or with mega-analyses? In short, are our findings the results of our methods rather than the psychobiology of the mind-brain? Questions like these have been inadequately considered by researchers in this field of research. This is the hole this book attempts to fill.

Obviously, the question of the role of meta-analyses is a very important one in today's cognitive neuroscience and increasingly so as the brain imaging methods have proliferated. It is the goal of this book to carefully consider the method from many points of view: its history, its methods, its limitations, its accomplishments, but, most of all, the empirical facts as we currently know them. In the following sections of this chapter, I further develop some of the topics introduced in the preceding narrative.

1.2 Why Do a Meta-analysis?

In the last decade there has been a substantial increase in the application of meta-analytic techniques to the evaluation of brain imaging data. Why this should be the case is immediately evident on examination of the current corpus of empirical data. To put it most baldly, much of modern work comparing brain images and cognitive processes is characterized by a lack of replication and a substantial degree of variability and inconsistency. The task of associating particular brain locations (or systems of brain locations) with particular cognitive process has turned out to be a much more difficult, if not intractable, problem than had originally been thought. Empirical findings vary for even the best-controlled and best-designed experiments; unfortunately, not all brain imaging experiments are well designed, nor are the independent variables adequately controlled.

Pooling the outcomes of many studies, in other words, *may* help us to discriminate weak and incorrect interpretations from compelling and correct ones. Pooling of data *might* allow us to filter out the extravagant and overhyped from the solid and robust scientific conclusions. Beyond the ability to filter out the outliers and converge on the most likely answer to some research question, it *may* also be possible to show that there is no robust support for certain cognitive neuroscience theories.

The premise on which the meta-analysis approach is based is that by pooling the data from a number of experiments, in the tradition of standard statistical practices, there should be some degree of convergence of the variable data onto some kind of an agreement concerning which brain mechanisms should eventually be associated with which kinds of cognitive activity.

A number of other more specific goals of meta-analytic techniques have been proposed; however, it bears emphatic reemphasis that the basic need for data pooling of this kind arises from the fundamental variability of the data forthcoming from experiments designed to relate cognitive processes and brain images. Image data pooling is supposed to be analogous to any other statistical analysis in which enlarging the sample size supposedly will progressively lead to an increase in the credibility of the outcome and the precision of measurement of a given experiment.

Cognitive neuroscientists (Kober & Wager, 2010; Wager, Lindquist, Nichols, Kober, & Van Snellenberg, 2009) have emphasized two fundamental goals to be achieved in a brain image meta-analysis: (1) evaluating the consistency of activations and (2) evaluating the specificity of those activations with regard to a particular cognitive process.

There is, however, another implicit role that makes the a priori error of assuming that specificity of localization is necessarily to be found in this kind of research. It seems to me that those who choose to assert localization a priori, as the following two articles did, are incorrect:

A goal of brain mapping in healthy subjects is to associate mental functions with specific brain locations (Costafreda, 2009, p. 33)

or

The goal of the analyses described subsequently is to localize consistently activated regions (if any exist[10]) in a set of studies related to the same psychological state (Wager, Lindquist, & Kaplan, 2007, p. 152)

Comments such as these are actually prejudging the outcome of the whole enterprise, specifically the answer to the localization versus distribution controversy. Not only may these investigators be begging the question of localization, but they may also be compounding the error of operating at the wrong level of analysis (macroscopic rather than microscopic). It is still a very contentious issue whether or not macroscopic brain regions are associated with cognitive activity in any simple way. It is more likely, as noted earlier, that the ultimate source of our "mental activities" is to be found in the microscopic activities and interaction of the innumerable neurons that make up the great networks of the brain. It is not at all clear, despite the enormous amount of research in this field, that the macroscopic activities of chunks of the brain will ever help us to understand how cognitive processes emerge from brain mechanisms.

This issue, nevertheless, is at its roots an empirical question requiring whatever tools we can bring to bear on the question. Because the system is intrinsically noisy, it is important as well as necessary to use all possible analytic tools to distinguish between signal and noise. In principle, meta-analyses might provide a means of extracting the signal (i.e., a representative brain activity) from irrelevant background activations. Whether or not they can do so in practice remains to be seen.

To the degree that the data are found not to be consistent, it is not only the details of putative cognitive modularity and brain localization that must be challenged, but also a much more far-reaching issue: Is cognition so broadly distributed that the whole search for localized brain correlates or biomarkers of cognitive processes is a bad question? Meta-analyses, either supportive or critical, may be the best way to answer this question.

Another important role for meta-analyses is that of dimension reduction (Tian, 2010). Data from imaging experiments are not only abundant but also multidimensional. Tian points out that there are a huge number of measurements made during each imaging session, and many of these measurements are needed to represent the activity of a single pixel. He notes, further, that analyzing data of this kind is computationally intensive but that this problem can be partially ameliorated by reducing the number of dimensions in various ways—meta-analysis being one of them. Therefore, meta-analyses or data pooling is, at the least, able to consolidate what is becoming an overwhelming mass of scientific findings, many of which produce inconsistent or contradictory outcomes, into a convenient and economical representation.

Meta-analyses can also potentially help us to determine how many brain regions are interconnected, "co-activated," or "concordant" (Wager et al., 2009). That is, even if it turns out that the brain responses to the simplest possible stimulus are complicated and distributed, some insight might emerge that will help us to understand the macroscopic organization of the brain. Given the high level of complexity of the interconnecting tracts pictured by diffusion tensor–type imaging procedures and other methods sensitive to the anatomy of the brain, this could conceivably provide a means of unraveling the complex pattern of interconnections among the parts of the brain. A remarkable new anatomical development in this context is the work of Modha and Singh (2010) in which the extensive interconnections among the parts of the Macaque brain were organized and displayed.

The variability and inconsistency that are so pervasively observed in this field of research result in large part from the empirical failure of individual experiments to exhibit sufficient statistical power. The key assumption underlying all meta-analyses, however, is that a number of statistically insignificant studies (due to their low power) may become collectively significant when their findings are combined into a large virtual sample. As noted, this is a hope that has not yet been fulfilled. In sum, meta-analyses may be very useful, but they still represent a controversial method for studying noisy and variable systems.

1.3 A Brief History of Meta-analysis

The idea that one could learn more by formally pooling the results of many inadequate or low-power studies than from individual studies lies at the heart of the meta-analysis approach. Indeed, the most fundamental postulate of meta-analysis is inherent in any statistical summary of data. From the earliest times when the personnel equation was first proposed by Friedrich Bessel (1784–1846) to explain discrepancies in astronomical observations, the idea that we could combine or pool the data from different observers (even if they were already averaged for each individual) was appreciated. However, one additional logical leap had to be made to develop fully the modern idea of data pooling. That leap was apparently first made by Simpson and Pearson (1904) in their pioneering meta-analysis of what then appeared to be a number of inconsistent results concerning the effectiveness of smallpox vaccinations. Their specific contribution was the central idea that the results of low-power experiments could be pooled to produce a virtual high-power experiment. By pooling the results of a number of different experiments, they made an important contribution to world health with their support of the efficacy of the vaccine.

A major next step in the development of early meta-analysis techniques was the pooling of probability values (p) of a number of experiments to get a composite p, a task first formalized by Fisher (1932) in his "inverse chi-squared" method. From that time, meta-analyses have been frequently used in a host of medical research topics, especially in situations in which only small differences between two treatments were observed.

The history of meta-analyses in the psychological and cognitive neuroscience fields, however, is much shorter. The first major historical step in the application of meta-analyses to the social sciences has been attributed to an Arizona State University Professor, Gene V. Glass (1976). Glass has also been credited as probably being the first person to coin the term "meta-analysis."

The story of Glass's contribution is quite interesting; it arose out a very specific controversy between Glass and a well-known psychologist, Hans J. Eysenck. Eysenck (1952) had been severely critical of the methods and techniques of clinical psychotherapy. He argued strongly that such "talk" therapies had little if any therapeutic effect beyond that expected from a placebo. The research supporting Eysenck's argument had come from what we would now consider to be a narrative analysis drawn from 24 studies involving 7,293 patients. Of these 24 studies, 5 used outpatient psychoanalytic therapies, and 19 studies used subjects who were provided with either eclectic custodial care or not treated at all. The results were not encouraging; only 44% of the psychotherapy-treated patients were "improved," whereas 64% of the eclectically treated or untreated patients were either "cured" or "improved." On this basis, Eysenck rejected at least the psychoanalytic version of psychotherapy as being effective.

Glass provides an interesting history of his response to Eysenck's critique in an unpublished conference paper entitled "Meta-analysis at 25" accessible at http://www.gvglass.info/papers/meta25.html. For reasons that were in part personal and in part professional, Glass describes how he challenged Eysenck in an effort to prove that psychotherapy actually worked. In 1977, Smith and Glass published an article in which they meta-analyzed the experimental literature (burgeoning by then to 375 relevant studies) on the effectiveness of psychoanalytic therapy. Based on this meta-analysis, Smith and Glass reported that they believed Eysenck was wrong—that psychotherapy did do better than the eclectic treatments. Their study was one of the first to point out that the particular style of treatment was not important, an idea that has been reinforced by recent research. Although the debate continues, whatever one concludes on the efficacy issue itself, it now seems most certain that no particular method of psychotherapy has any advantage over any other with the possible exception of a slight advantage attributable to behavioral methods. This conclusion has been substantiated by a number of studies including Smith and Glass (1977), Shapiro and Shapiro (1982), Lipsey and Wilson (1993), and Anonymous (2004).

Glass's (1976) report also described some of the earlier work that had been done in psychology in which data were pooled from a number of what were often inconsistent findings. Some investigators (e.g., Maccoby & Jacklin, 1974) simply used narrative techniques to assert what they thought was the joint impact of as much of the experimental literature as they could collect. Others, however, began to use techniques that are more formal. Astin and Ross (1960), Schramm (1962), Dunkin and Biddle (1974), and Sudman and Bradburn (1973) all used simple "voting" procedures (i.e., comparing the relative number of significant and insignificant outcomes) on which to base their conclusions. Although relatively simple in terms of their methodology, these voting techniques must certainly be classified as an early form of meta-analyses.

Glass's (1976) seminal article proved to be the modern starting point for the formal process that he named meta-analysis. The history in the 1980s was characterized by an increasing number of reports including those by Shapiro and Shapiro (1982), Bond and Titus (1983), Guzzo, Jette, and Katzell (1985), Hyde and Linn (1988), and Bornstein (1989). In the next decade, the technique was used by an increasing number of investigators. Rosenthal and DiMatteo (2001) tabulated 24 major studies (mainly from the 1990s) in which problems in behavioral medicine, social and clinical psychology, as well as organizational psychology, were meta-analyzed.

During the entire 35-year history since Glass's introduction, naming, and formalization of the method, there have also been continued efforts to standardize the field. Texts such as Cooper and Hedges (1994) and their updated handbook (Cooper, Hedges, & Valentine, 2009) provide advice on how to properly carry out meta-analysis and avoid some of the biasing pitfalls that we consider in chapter 2.

Despite these positive attempts to standardize the field, the use of meta-analysis in psychology has not lacked for criticism. Shercliffe, Stahl, and Tuttle (2009) have been among the most vociferous critics of the entire approach to pooling data. They challenged the whole idea that meta-analysis "is an inherently superior technique as compared to other forms of review" (p. 413). Specifically, they pointed out that most of the purely psychological meta-analytic studies they examined were characterized by ". . . inconsistent/incomplete sampling procedures, a lack of reporting of even the basic methodological information (thus making replication difficult), and a tendency not to address issues that may affect the validity of the results" (p. 424). They went on to argue: ". . . the conclusions of these meta-analyses are influenced to some degree by private judgment, collective views, and personal style, which is, notably, the same criticism that Glass et al. (1981) made of narrative reviews" (p. 424).

Shercliffe, Stahl, and Tuttle also pointed out that psychological meta-analyses, to their detriment, depend heavily on the prevailing statistical Zeitgeist of modern psychology with its possibly misleading emphasis on significance testing.[11] They suggested, further, that many personal decisions are made in the execution of a meta-analysis that do not differ in

Meta-analysis: The Idea

kind from those made in a narrative interpretation. If one extrapolates from their purely psychological studies to the neuroscientific ones that are the topic of this present book, meta-analyses, likewise, may not be the elixir that we are seeking in our search for the answers to questions about brain-cognition relationships. Without doubt, valid or not, meta-analyses are being used by an increasing number of psychologists to deal with the large number of reports of comparable psychological functions. The need is great. As Glass (1976) so presciently pointed out: "We face an abundance of information. Our problem is to find knowledge in that information. We need methods for orderly summation of studies so that knowledge can be extracted from the myriad individual researches" (p. 4).

The current breadth of application of the meta-analysis method to psychology is but barely hinted at by a Google search for the terms "meta-analysis" and "psychology," a search that turned up about 1,420,000 results. Other researchers have suggested that there are hundreds of psychological articles in which the term "meta-analysis" is used in the title published each year. Although there are many duplications and uncertain references, it is currently beyond doubt that meta-analysis has become a very pervasive technique for the analysis of psychological data.

Now, a new role has emerged for meta-analyses. In addition to purely psychological studies in which standard descriptive and inferential statistics have a long history, meta-analyses of neuroscientific findings are becoming popular. Once again, the need is based on the considerable inconsistency observed among the findings obtained from different experiments. Indeed, the reasons for applying meta-analyses to neuroscientific data are fundamentally no different from those required in any other science displaying considerable variability.

Among the first reported applications of a meta-analysis to brain image data was a study by Tulving, Kapur, Craik, Moscovitch, and Houle (1994) on hemispheric representation of encoding and retrieval processes as reflected in PET scans. Their meta-analysis was a fairly simple one typical of those in use at the time; they simply tabulated which side of the brain was active in the encoding and retrieval processes, respectively, in a series of 14 experiments using PET scans. In this and some other early attempts to pool data, a meta-analysis simply meant comparing or tabulating the results from a number of experiments in what was little more than a simple voting procedure. The results of their study were startling in their simplicity; all encoding activations occurred in the left hemisphere and all retrieval activation in the right hemisphere.

More formal meta-analyses were subsequently carried out on PET imaging data by Paus (1996) and Picard and Strick (1996) in which statistical measures such as means and variances were computed to support narrative judgments or simple votes. A more detailed history of these relatively early research projects can be found in an essay by Fox, Parsons, and Lancaster (1998).

It is not obvious where the first applications of meta-analysis were made to fMRI data; however, it is clear that by the 1990s many investigators were concerned with merging the highly variable brain image data obtained from different runs within an experiment or between different experiments. One of the first discussions of the problem at that time was presented by Robb and Hanson (1991). Fox and Woldorff (1994) also discussed the issue shortly after that; a main interest of the latter investigators was in normalizing the different spatial brain maps obtained from different subjects. They briefly alluded to the idea of meta-analysis as a potential means of "integrating maps among subjects" without expanding on how this might be done. To the best of my knowledge, the first experiment in which the term meta-analysis was explicitly used in an fMRI-based experiment was reported by Videbech (1997). Videbech was interested in anatomical changes that could be associated with affective disorders—itself a very wide-ranging spectrum of psychological dysfunctions. The "analysis" in this case was a more formal analysis based on the Mantel-Haenszel statistics test, a test of the null hypothesis for a number of categories or experimental findings.

By the late 1990s, investigators such as Fox, Parsons, and Lancaster (1998), were championing the use of meta-analysis for overcoming many of the problems engendered by the data variability observed in PET images. By this time, the need and use of meta-analyses greatly expanded as the number of similar brain imaging experiments exploded. Cabeza and Nyberg (2000), whose graphical meta-analysis is still one of the most influential sources of information in this entire field of inquiry, pooled the results of 275 PET and fMRI experiments from a broad range of cognitive topics. (Most of the reports used in their analysis were probably based on PET images.) When Phan, Wager, Taylor, and Liberzon (2002) published their meta-analysis of brain responses to emotional stimuli, they could call up 12 fMRI studies as well as 43 PET reports for this narrow topic alone. The floodgates have been open for the last decade. Our continuing discussion considers the work in its technical context rather than the historical one characterizing this brief introduction.

1.4 Some Relevant but Basic Statistics

The decision problem in interpreting and understanding brain images is comparable to any other research that is characterized by variable or noisy data. It is necessary to apply statistical methods to determine the probability of the truth or falsity of a hypothesis. There are no absolute answers because the same value of an effect can often be attributed to the signal-plus-noise or the noise-alone distributions. In the brain-imaging context, this can lead to two distinct types of errors. A type II error—a miss—would have been made if the investigator concluded that there was no difference between the responses to control and experimental conditions when there actually was. A type I error in brain imaging—a false alarm—would be equivalent to reporting as an activation an event that was actually the result of noise alone.

Lieberman and Cunningham (2009) pointed out that in our efforts to avoid false alarms or type I errors in recent research using brain images, there has been an exacerbation of type II errors in which a real difference between the experimental and the control condition was missed. This is an inevitable result of the criterion level at which one decides that there is or is not a significant response present in the image. As hits go up, misses must go down. This situation, therefore, leads to a conservative approach to determining the presence or absence of real activation areas. Whether or not this is a desirable development is a judgment call that, like all other statistical decision judgments, must be based on arbitrary criteria such as the relative costs of hits and misses.

An arbitrary criterion value demarcates between what we will accept as noise and what we will accept as signal plus noise. It does not exactly determine the presence of a signal, only the probability that the signal is likely to be present. If the criterion value is set too low, then a larger number of false-positive activations will be associated with the experimental condition. If the criterion is set too high, then there will be many activations overlooked that should not have been. The important point is that there is a balance between false alarms and misses (i.e., between type I and type II errors) that it is impossible to overcome. This basic situation shows up in even the quantitative methods.

To put it most succinctly, the purpose of a meta-analysis is to determine the probability that a distinctive brain image actually is correlated or associated with a cognitive process by pooling data from a number of experiments. This increases the power of an analysis by building an adequate virtual sample size from a number of experiments that individually had inadequate power.

Sample size is a major factor in determining the robustness of a judgment that there is or is not a signal buried in the noise. So, too, of course, is the magnitude of the effect being measured. There is little that can be done to manipulate the size of the effect—that is what is being dependently measured. However, increasing the number of subjects in an analysis can determine whether the measured effect is going to be of sufficient power to exceed some criterion of significance and, thus, to be accepted or rejected as a real response—a signal.

No matter what measure or criterion is used to determine the significance of the outcome of an experiment, sample size plays a central role in determining its impact, value, or utility. Remembering that the data themselves (i.e., the effects) are going to be the primary criterion of significance, sample size appears throughout any statistical analysis as a controllable feature.

Sample size may also be considered from another point of view—its role in enhancing the signal-to-noise ratio. As more and more data are accumulated from larger groups of subjects, repetitive signals will be reinforced, and the noise, if random both in a negative and positive direction, will tend to cancel out. This is the basis of statistical averaging as a means of extracting a signal from a complex noisy background—a process that has been widely used to extract the event-related potential (ERP) recorded with EEG-like methods. Averaging of this kind is becoming more frequently applied to newer high-speed techniques such as event-related fMRI (efMRI)

In short, the problem is that the determination of a significant association between brain responses and a cognitive process is not a simple yes-no type of situation in which there are clear-cut distinctions to be drawn between conditions, from subject to subject, from experiment to experiment, or between effect distributions and shapes. Instead, it represents a

Meta-analysis: The Idea

complex analytic situation replete with inconsistent and noisy responses that are either quasi-random or driven by unknown forces in a way that makes them appear to be nearly random. The very same brain response may, in one instance, reflect an activity associated with a cognitive process and, in another, be nothing but noise.

The variability in activation areas can be so great that we usually must make arbitrary decisions about the reality and relation of the neural and the cognitive processes based on probabilistic analyses rather than deterministic ones. Although there is little that can be done about the intrinsic noise in such a situation—noise is a function of the system being studied and the methods used to measure it—it is possible to increase the probability of not making a type II error by increasing the power of an experiment, and this effectively means increasing the sample size. Achieving this ideal goal is the essence of why one would want to use the analytic methods we call meta-analyses.

In sum, it was argued by Glass and others: "Why not pool the results from many experiments to increase the expected power of a study?" Whereas an individual experiment cannot always afford to use enough subjects, the assertion is that it might be collectively possible to improve the power of our analysis by combining the data, creating a large "virtual" sample size, and thus reducing the possibility of a type II error when we draw our conclusions. Unfortunately, the process of pooling data from different experiments is more difficult than the simple rhetorical question implies. The basic reason for this difficulty is the variability of the outcomes of even the best-designed experiments.

1.5 Other Problems with Data Pooling

From the time of the earliest studies of localization using functional brain imaging, there has been concern that the simplistic association of a particular brain region with a particular cognitive process was flawed in some fundamental way. For example, although speech has long been specifically associated with Broca's area, it is not well known that there was a considerable argument against such a narrow localization based on the variability of individual responses and distributions. (This controversy was extensively discussed in Critchley & Critchley, 1998.) The problem was that the way in which individual responses were pooled or combined to give an "average" location (e.g., Broca's area) did not reflect the fact that there were vast individual differences among individual cases or that Broca's area was far more extensive than it had originally seemed. On this theory, a narrowly localized "Broca's area" was a statistical fluke representing a mythical average but was not neurophysiologically meaningful as a function-specific "center" that uniquely instantiated speech in individuals.

Nevertheless, throughout the twentieth century, research was in large part aimed at specifying as narrow a localization as possible by averaging individual activations. Relatively few investigators examined the individual records to estimate the intersubject variability. Although the individual records were slightly different because of both vagaries in experimental design and neuroanatomical noise, it was implicitly assumed that appropriate statistical manipulations could infer what was the standard brain location for a given cognitive process.

In a typical psychological experiment, means and standard deviations are computed from numerical values of dependent variables that are defined along well-defined metric dimensions such as effect size or reaction time. (The metric of a dimension is the geometric function that defines the distances between pairs of points in a space.) Many psychological dimensions (such as reaction time) have regular and equal intervals between any pairs of points along that dimension because they are linked to physical time. Other kinds of measurement may have unequal but still regularly defined distances between pairs of points. An example is the system of units produced by subjective magnitude estimates that seem to be best fit by exponential functions. It is relatively easy to deal with either of these two cases: As more and more data are accumulated, these comparable values can be further pooled to give ever better estimates of the metric (and other measures) of the dimension under examination. Statistical distributions for these kinds of metrics are clearly defined and can provide precise and meaningful estimates of the variability and central tendency of the specific parameter under examination.

However, it is extremely difficult to try to average responses that have no underlying metric but in which intervals are irregular and unequal. The absence of a regular metric challenges the whole idea of combining data by any kind of an averaging process and even raises questions about the most basic postulate of the quantifiability of cognitive responses.[12]

Thus, although we may have obtained a set of values from different runs that can be numerically combined to give a seemingly quantitative estimate of an average value or variability along a dimension, these pooled numbers may be meaningless. We may be combining values for which there is no stable underlying metric of the distance between spatial pairs of points. It would be as if we had an elastic ruler that constantly shifted its units of distance from one measurement to another.

In the absence of a dependable and solid metric, the variability situation in the case of brain images, defined as they are along spatial dimensions, is considerably more complicated than in the usual psychological experiment. In a typical imaging experiment (many specific examples of which we will examine later in this book), the activated regions for different pools of subjects may not even overlap. For many of the reasons discussed earlier, distinctly different locales may be illuminated or contrasted by slight variations in stimulus conditions.

If we consider the simplest case in which two non-overlapping regions are activated in two different subjects, what does a spatial "average" of the two mean. Does this bare-bones response pattern mean that the two spatial extents represented by the two brain images can be averaged to give a meaningful estimate of an intermediate region within which neither one showed any activation? Does it mean that only the common regions are to be used to define a common activation region? The same answer applies to both of these rhetorical questions—of course not. Although we can carry out the mathematics of the spatial averaging, the "averaged," intermediate location is nonsensical without a persistent and stable ruler.

Nor would spatial averaging always be meaningful even if we had a good metric of brain space, which does seem to be irregular. As a simple analogy, consider the following. New York and Los Angeles, on the average, lie somewhere in the Midwest of the United States, but neither is in that "average" geographical location, and nothing is added by adding some place in Nebraska to our knowledge of the locations of either New York or Los Angeles. Such pseudo-averaging of spatial locations must be interpreted in quite a different manner. There are perfectly sound reasons why two distant brain locations may have developed, as they did, not the least of which is that the whole brain is responding more as a unit than in some separable manner. Yet, as we see in chapter 3 when we examine some of the data on individual differences, averaging human brain images often leads to totally nonsensical as well as highly variable responses that preclude localized assignment of a cognitive process to any circumscribed brain region.

Reliability means something quite different in this case than in those that permit averaging of quantifiable values along a single metric dimension such as the distribution of heights in males. This kind of interpretive error was exemplified in the recent statistical analysis of the popularity of President Barack Obama. It was repeatedly stated that he had an "average" approval rating of around "50%," suggesting that most people were more or less neutral with regard to their evaluation of his popularity. Although this number seemed to mean something, on closer examination, it turns out to be a meaningless number. The actual bimodal distribution of the responses to this survey showed that there was a very large number of people who strongly disapproved of him and a very large number who equally strongly approved. The middle range was almost completely unpopulated. The "50%" value thus has little value as a measure of his popularity. It is an "average" of the "unaverageable."

Similarly, there may be perfectly good reasons why two extensive but non-overlapping brain regions should not be averaged. One important reason is that they may be two interconnected portions of a broadly distributed system. Averaging the two regions would thus attach special significance to the common parts of continuous but imaginary localized regions that actually did not exist in any functional sense. This common erroneous logic pervades any attempt to draw conclusions from only slightly overlapping brain regions. This may be referred to as the "Venn fallacy." Figure 1.1 shows the logic behind this common error.

The Venn logical approach, common in this field, argues, for example, that if three experiments aimed at answering a brain localization question jointly produced a single overlapping activation pattern, the common region (or "union") actually defines the relevant and narrowly circumscribed brain locus. That is, the incorrect conclusion drawn is that it is only the small common region that is presumed to be encoding the manipulated cognitive activity. However, such logic ignores the possibility that the entire extent of all three broader regional distributions may also be real, true, or valid neural responses. To the contrary, the union is a fiction arising out of the desire to narrowly localize whatever cognitive

Meta-analysis: The Idea

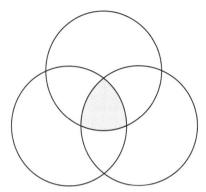

Figure 1.1
The illogic of the Venn fallacy in which noncommon areas are ignored.

process is being studied in the brain. There is, from this latter perspective, nothing particularly compelling about the accident that the three broadly responding regions shared a common field or union; it is our cultural scientific proclivity to search for narrow localization, not a compelling set of data, that drives us to make the Venn fallacy. Happily, broad distribution and multiply interconnected nodes devoid of functional specificity now seem to be offering an alternative metaphoric concept of brain and cognition.

Furthermore, the initial appearance of two or more separate regions may itself be an artifact. Regions may appear to be separate because the threshold was set too high by the imaging investigator. High thresholds produce a few narrow peaks; low thresholds produce diffuse activity distributed over more of the entire brain. It is entirely possible that what was in actuality a very broad, brain-wide activation pattern was fallaciously converted into isolated peaks by a high threshold.

The central issue, therefore, is the variability of the brain responses from individual subjects—the non-overlapping portions of the activation regions may represent neurobiological responses that are as real as any others and not just the random effects of some kind of neurobiological noise. All too often, we simply throw out the noise in our effort to find localized regions of cognitive significance.

Despite the importance of variability in many aspects of cognitive neurosciences, variability remains a relatively underdeveloped interest among most cognitive neuroscientists. Among the few who have tackled what I believe is the central core of the brain imaging–cognitive comparisons are Bennett and Miller (2010). They have been leaders in exploring central questions such as "How reliable are the results from functional magnetic resonance imaging?" They emphasize that measuring variability in brain imaging experiments is not necessarily a direct or easy task. All of the methods for doing so are limited by constraints of one kind or another; however, among the best measures of variability they cited was the intraclass correlation coefficient (ICC) defined by the following equation:

$$C = \frac{\sigma^2_{between}}{\sigma^2_{between} + \sigma^2_{within}} \tag{1.1}$$

Where $\sigma^2_{between}$ is the variance of interest or the spread of scores around a grand mean of all groups, and σ^2_{within} is the variance or the spread of scores within groups. The symbol σ^2_{within} can also be interpreted as the noise in the system; to the extent it is large, ICC scores are low, and vice versa. This provides a range of values for variability that goes from perfect association (1.0) to none (0.0). Other methods described by Bennett and Miller, less frequently used, have included maximum likelihood estimates and the coefficient of variation.

Bennett and Miller (2010) sum up the current situation concerning reliability as follows.

There is little agreement regarding the true reliability of fMRI results. While we mention this as a final conclusion from the literature review, it is perhaps the most important point. Some studies have estimated the reliability of fMRI data to be quite high, or even close to perfect for some tasks and brain regions (Aron et al., 2006; Maldjian et al., 2002; Raemaekers et al., 2007). Other studies have been less enthusiastic, showing fMRI reliability to be relatively low (Duncan et al., 2009; Rau et al., 2007). Across the survey of fMRI test-retest reliability we found that the average ICC value was 0.50 and the average cluster overlap value was 29% of voxels (Dice overlap = 0.45, Jaccard overlap = 0.29). This represents an average across many different cognitive tasks, fMRI experimental designs, test-retest time periods, and other variables. While these numbers may not be representative of any one experiment, they do provide an effective overview of fMRI reliability. (p. 145)

We thus are left with two important but unanswered questions:

1. How variable are the responses from comparable studies?
2. Does the observed variability preclude robust associations between cognitive and brain image processes?

These questions arise repeatedly throughout this book, especially in chapter 3, where I carry out actual comparisons of data. For the moment, we have to consider the reliability question to be unresolved. However, it will become clear from our discussions that reliability is quite low when one compares data from all levels of analysis.

Bennett and Miller (2010) also reported the results of a meta-analysis of 15 studies in which the ICC measure was used to measure reliability. Values ranged from essentially no correlation (0.0) to nearly perfect correlation with a wide scatter of intermediate values. Other methods produced comparable variations in the variability scores. Significantly, we see later that the higher values they reported do not occur when one compares meta-analyses rather than individual studies.

Bennett and Miller also considered the question, "What level of reliability should be considered to be acceptable?" In lieu of a specific answer to this question, they pointed out that there are a number of reasons why such a question cannot be answered. Equipment, subject samples, individual differences, and laboratory protocols all influence the reliability measures. However, there are two reasons that seem to me to stand above the rest: (1) the intrinsic arbitrariness of the cognitive task or process created by the poor definition of psychological constructs for which a neural foundation is being sought,[13] and (2) the neurobiological variability both within the individual brain and between brains. The resulting inconsistency of the raw data resulting from these two factors argues strongly that the variability of cognitive neuroscience findings has been greatly underestimated. As a result, the theoretical significance of brain imaging findings has been seriously overestimated.

Clearly, the problem of reliability and, therefore, the acceptability and meaningfulness of such variable and unreliable data typically encountered in brain imaging studies are going to be with us for a long time. Equally clearly, much of the current crop of brain imaging research is sufficiently variable as to possibly preclude any simple answer to the question of how the brain represents or encodes cognitive functions. Many current experiments associating brain locales and cognitive processes do not replicate, as extensively discussed later in this book and in my earlier works. Not only do we inappropriately pool variable individual response patterns, but also the pooled meta-analyzed estimates fail to converge on simple answers to the complex problems; as a result, they may actually impede our efforts to unravel the great mind-brain problem.

Bennett and Miller (2010) summarized their concern about the reliability of brain images by pointing out that the problem has not enjoyed the attention it deserves. Despite the fact that there are a number of different methods for measuring reliability, the cognitive neuroscientific community has not converged onto a generally acceptable level comparable to the .05 criterion for statistical significance in conventional psychological research.[14]

It is the purpose of this book to dig deeper into the problem posed by the variability of brain imaging data as used by cognitive neuroscientists at several different levels of analysis. This includes the comparison of meta-analyses, a task I refer to as meta-meta-analysis. This first chapter introduces the problems and sets the stage for the subsequent chapters.

Chapter 2 is aimed at specifically exploring the meta-analysis methodology itself. In it, I examine the kinds of meta-analyses that have been used to evaluate pools of experimental results. I discuss the advantages and disadvantages of the various methods, particularly with regard to the many sources of bias that can distort our understanding of the pooling process.

Chapter 3, the heart of this book, discusses and compares the empirical findings that have been forthcoming from the application of brain imaging techniques to the study of cognition. Comparisons are carried out at all levels of data pooling.

Chapter 4 examines the nature of the theories and interpretations that have emerged from the huge mass of data that has become available in the last two decades since the invention of functional MRI. It also discusses the role of brain imaging in theory development.

Chapter 5 draws conclusions from current empirical results and identifies a number of emerging principles that help to encapsulate the current state of the art in cognitive neuroscience.

2 Meta-analysis: The Methodology

2.1 Introduction

In chapter 1, I discussed why there is such an extensive and obvious need for meta-analyses of brain image data. The nutshell expression of this need is that many, if not most, individual experiments do not have sufficient statistical power (largely because of inadequate subject sample sizes) to distinguish between random artifacts (the null hypothesis) and actual neural correlates (significant differences). One result of their low power is that the outcomes of some experiments purporting to study the same phenomenon are often very variable. The proposed solution to this problem is to build a large "virtual" database by pooling the results of a number of experiments—in other words, to carry out a meta-analysis.

As easy as it is to express the putative solution inherent in a meta-analysis, there are many difficulties in carrying out such a procedure. Each of the experiments in the pool must be sufficiently similar to the others in terms of their stimuli and effects to justify combing them. In a world of poorly defined cognitive processes, procedural variations, and what appears to be a substantial amount of neurobiological variability among people, this is not an easy task. Nevertheless, a substantial number of cognitive neuroscience meta-analyses have been carried out in the past decade as the idea has become increasingly popular.

A typical meta-analysis often involves complex statistical manipulations of the data. Therefore, the possibility that the statistical manipulations themselves may introduce artifacts (false associations of brain regions and cognitive processes) should not be overlooked. Indeed, it is not necessarily preordained that individual brains are similar enough in terms of their representational mechanisms to be compared or pooled. Nor is our technology sufficiently robust to guarantee that the answers to our research questions can or will be answered by the best-intended efforts to increase experimental power. In systems with a great deal of variability, conclusions are always based on probabilities. Both type I (false alarms) and type II (misses) errors are possible no matter how much data are pooled. As a result, it is necessary for investigators to be especially watchful when drawing conclusions about the findings from meta-analyses. Whatever interpretations and theories we may ultimately generate depend on the reliability of our empirical data. We will see later in this book how that requisite empirical consistency is remarkably hard to find. Thus, it is important that we also consider the sources of bias, error, and variability that seem to plague complex meta-analyses of brain imaging data, not the least of which is the powerful force of the current Zeitgeist that often champions one idea or conclusion over another, sometimes even before an experiment or analysis is carried out. The prevailing Zeitgeist also tends to make us less open to new interpretations than is desirable in a dynamic science. It is surprising how powerful the Zeitgeist can be even when a substantial amount of new data actually supports a different view than the currently popular one.

This chapter has two goals. The first is to classify in at least a preliminary manner the kinds of meta-analyses that have been developed in recent years. The second is to tabulate and discuss the sources of error, bias, and inconsistency characterizing the current experimental literature. Our main concern throughout this chapter is with the spatial patterns obtained with brain imaging devices. Because the preponderance of results in this literature are presented in terms of spatial localization, we must start by introducing the nomenclature that is used to specify location in and on the brain.

2.2 The Language of "Where" in the Brain

Consider, for a moment, the difficulty inherent in specifying the answer to the superficially simple "where" question. Because brains vary in size and, to a degree, even in shape, it has been considered essential to develop a standardized nomenclature to supplement qualitative pictures. Although there have been many efforts to standardize the nomenclature used to define brain locations, none is definitive. Three different reporting modes are typically used to designate the extent of the responding activation areas.

1. Brodmann areas
2. Narrative "relative" locations
3. The Talairach-Tournoux system

The traditional system has been for many years the Brodmann areas (BAs) proposed in 1908. Despite its longevity, the BA system is not without its flaws. It was originally charted based on the anatomical shapes of neurons (cytoarchitectonics) that Brodmann observed microscopically in a single sample of human brain. Therefore, the boundaries of the BAs are vague and uncertain given both the variability of individual brain anatomy as well as Brodmann's ill-defined boundaries themselves. Furthermore, because there are about 50 BAs, they vary significantly in their extent and shape. Therefore, an activated fMRI response could overlap two BAs; in other cases, only a portion of one BA might be activated. Furthermore, BAs extend over broad regions of the brain in a way that could make any restricted localization by designating a particular BA misleading. In particular, as shown in figure 2.1, Brodmann's original areas 19, 10, and 21 were elongated areas that spanned extensive regions of the brain. Others such as BAs 41, 42, 43 are small and close together, and activation (depending on an arbitrary threshold) might spread or seem to spread across all three.

A second means of comparing brain locations is based on narrative language exemplified by such phraseology as "posterior middle temporal gyrus" or "orbitofrontal gyrus." The arbitrary and qualitative (as well as idiosyncratic way) in which this terminology is used, however, makes it difficult to compare locations from one study to another in a precise manner.

The third method of locating activation regions on the brain is the one proposed by Talairach and Tournoux (1988) and modified at the Montreal Neurological Institute (MNI) by Evans, Collins, and Milner (1992). These systems use a Cartesian three-dimensional coordinate system to define the locations of activations both within and on the surface of the brain. This new type of three-dimensional coordinate system is more and more often being used in brain imaging studies. However, the Talairach and Tournoux coordinate system represents each activation area as a point in space rather than as a distributed region. This can be a serious problem leading to an

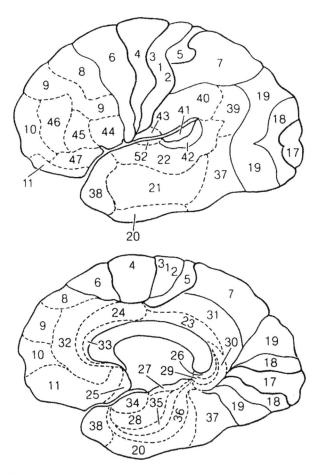

Figure 2.1
A modern version of Brodmann's map of brain regions. From Cabeza and Nyberg (2000) after H. C. Elliot's *Textbook of Neuroanatomy* (1907).

underestimation of the extent of a response because the activations are never restricted to a narrow point. Although this problem is partially compensated for by the use of such artifices as uniform Gaussian distributions mathematically defined around that point, even that method is not fully able to characterize the distribution of the activation areas; irregular regions of activity would not be adequately described by this method.

Furthermore, brain coordinate systems of the Talairach and Tournoux type (x, y, and z) are also difficult to compare quantitatively or to plot graphically. Many such attempts using three-dimensional plots (on two-dimensional paper) end up with an indiscriminable blob representing clusters of activation areas that are difficult to visualize. Figure 2.2A, for example, is a raw plot of the activation peaks associated with emotional stimuli reported by Murphy, Nimmo-Smith, and Lawrence (2003).

The two lower graphs, figure 2.2B and C, represent subsets of the responses shown in figure 2.2A for two different experimental conditions. Figure 2.2B represents activation peak responses for "positive-negative" valence emotions, and figure 2.2C depicts the responses to an "approach-withdrawal" condition. Despite the use of the black dots to distinguish between the activations forthcoming from each of the two conditions, it is difficult to distinguish visually between the two lower graphs.

The original Talairach and Tournoux atlas was based on the brain of a single female with an unusually small brain. Furthermore, their coordination with the BAs was based on an arbitrary inspection of this woman's brain and did not precisely relate to the traditional regions suggested by Brodmann. For reasons such as these, more precise standard brain maps such as the MNI atlas of the brain have become widely used in recent years. Neither, however, is completely representative of all brains, and the three-dimensional coordinates given for particular locations are subject to considerable error.

Specific transformations also have been suggested to convert Talairach and Tournoux coordinates to the MNI version (e.g., by Lacadie, Fulbright, Rajeevan, Constable, & Papademetris, 2008). Attempts to convert between the Montreal Neurological Institute and the Talairach-Tournoux systems continue to be major efforts in current cognitive neuroscience. Currently several types of mapping are utilized, unfortunately often without any indication of which one was actually used. The identification and measurement of comparable locations on the brain have become an increasingly

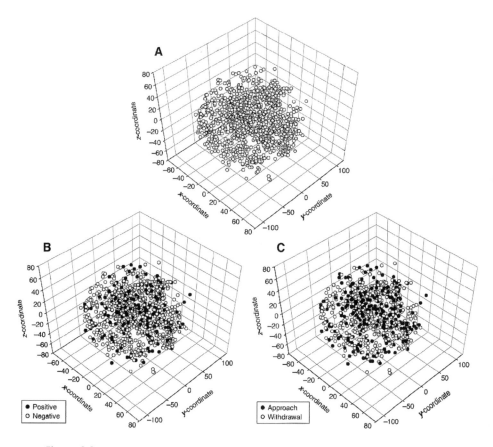

Figure 2.2
Talairach and Tournoux plots of all reported activation peaks for emotional stimuli. (A) All emotions; (B) positive/negative valence emotions; (C) approach/withdrawal emotions. From Murphy, Nimmo-Smith, and Lawrence (2003) with the permission of the Psychonomics Society.

difficult problem as the resolution of the MRI systems increases and voxel sizes decrease.

It has been suggested by other investigators that it should be possible in principle to interconvert from any one of these three types of systems to the other two. However, such an approach was always inhibited by the arbitrariness of the narrative localization system, the imprecision of the BAs, and the pointlike representations of the x, y, and z coordinates in Talairach and Tournoux systems. Indeed, the inadequate manner in which Talairach and Tournoux originally tried to rationalize their coordinate systems with the BA system has long been criticized.

2.3 A Typology of Meta-analyses

Although modern meta-analyses are often complex statistical manipulations, the basic idea characterizing them is also inherent in the most basic of data-pooling methods. Indeed, the whole idea of using multiple subjects in an experiment is operationally no different in principle from that underlying the most complex interexperiment, meta-analytic data pooling. In the single experiment, we pool the results of many subjects; in the meta-analysis, we pool the results of many experiments. The basic need for pooling or averaging individual subjects emerges from the indisputable fact that there are substantial differences among individual subjects (as documented in chapter 3). If one limited experiments to examining the results of single subjects, the degree of intersubject variability would produce chaos in the kind of cognitive neuroscience that we are discussing here.

In point of historical fact, single-subject experiments are not unusual. We call them case studies, and they are probably among the greatest source of bias, error, and flawed theoretical explanations throughout cognitive neuroscience. Often, the problem is exacerbated by injudicious selection of the results produced by a single subject from a larger sample that collectively did not otherwise support some general conclusion.[1]

The basic need for comparing different experiments, as noted earlier, arises from the fact that there are substantial differences between the outcomes of what are intended to be similar experiments. This problem is further exacerbated by the fact that it is expensive and time consuming to carry out fMRI brain imaging studies. As a result, not all reports are blessed with adequately large samples, and given their wide variability, their statistical power can be severely limited. Subject sample size in some tomographic methods is further limited because it is also currently considered unacceptably invasive to carry out PET studies for nontherapeutic reasons. The invasiveness involves the injection of radioisotopes. It is also necessary to have an onsite cyclotron. Therefore, the preponderance of experiments published in the last decade report the use of relatively few subjects and few trials on any single subject. Compared to the ease with which many subjects can be recruited and tested in a typical human perception experiment, this is a major impediment. Wager, Lindquist, Nichols, Kober, and Van Snellenberg (2009), for example, reported that in the 415 studies they reviewed only 4,856 subjects were used. This is an average of only 11 or 12 subjects per experiment, and some experiments used as few as 4 or 5 subjects. It is problematic, depending on the variation among subjects, whether these small numbers are sufficient to support conclusions and theoretical syntheses.

Nevertheless, even for these small numbers and the resulting low power of their experimental protocols, experimenters often felt confident enough to publish their findings and presumably draw strong conclusions from their work. Given the unreliability of the typical brain image, this may have been a reckless strategy.

Because of the relatively low power of many cognitive brain imaging experiments, there have been many recent attempts to develop powerful methods to further pool the outcome of a number of experiments. The goal is to enhance the power and to determine what, if any, truly robust associations can be made between brain activations and cognitive states.

Regardless of the method, it must also be reiterated that the goal of almost all brain image–cognitive process comparisons in cognitive neuroscience is to answer the "where" question. That is, at the most fundamental level, brain imaging experiments can only tell us *where* something is happening in the brain and not *how* cognitive processes are produced by these macroscopic measures. Ignoring for the moment that the "where" question may be a bad one, the main thrust in any brain imaging study is to support the idea that there exists more or less localized activity corresponding to particular cognitive functions. How local is defined in this context may be quite variable; however, in principle, even the most distributed system answer to this question is just a question of how much and the degree to which localization exists.

Given the variability of the results from individual subjects and from individual experiments, it is always difficult to answer the "where" question with precision. The hope is that by a pooling or averaging of what appears to be disparate or inconsistent data, we may be able to zero in on brain regions (or systems of regions) with specialized cognitive functions. However, this important and challenging task is loaded with sources of error and carries a considerable baggage of noisy and erroneous data. Wager, Lindquist, and Kaplan (2007), for example, estimate that "17% of the total number of reported peaks" in a sample of "195 studies of long term memory" (p. 151) are false positives. Distinguishing between the false positives and valid cognitive-brain associations is a major goal of meta-analysis.

In this section, I present a description of the types of meta-analysis that have been used in recent years to pool the results of multiple experiments to increase their collective power. These include the following approaches:

- Inductive narrative comparisons
- Voting
- Conventional statistics (otherwise known as classic or Glassian meta-analysis)
- Pictorial and graphical superimposition
- ALE
- KDA and MKDA
- Replicator dynamics

Later sections of this chapter examine the sources of bias that continue to distort both the empirical outcome of this data pooling and the theoretical conclusions that arise from them.

2.3.1 Narrative Interpretations

Much of modern science does not use sophisticated analyses or even quantitative methods to interpret the collective meaning of its findings. In many sciences, it is simply assumed that an experienced and well-informed scholar has a powerful ability to perceive the patterns manifested in a complex, multidimensional database. Simple narrative descriptions based on informal inductive thinking are still a mainstay, especially in the social and behavioral sciences. There is, indeed, much evidence to support the contention that people are good at many kinds of inductive data pooling; human beings demonstrate wonderful abilities to recognize faces, to track complex movements, or to predict the outcome of subtly nonrandom sequences.

However, it is also well known that people tend to perceive patterns even in situations in which the sequences are actually quite irregular, even random. It is does not take too long observing gamblers to appreciate how self-deceptive people may be in "seeing" patterns that do not actually exist. For example, most people do not appreciate that "hot" streaks are much more consequences of statistical improbabilities than of any sequential causality. The ubiquity of human illogic has been repeatedly described by psychologists such as Tversky and Kahneman (see, for example, their now classic 1983 article as well as the article by Gilovich, Vallone, and Tversky [1985] on the "hot hand" in basketball).

Furthermore, what research we arbitrarily choose to include in our narratives can strongly influence the nature of our conclusions. Unfortunately, scholars and scientists are as likely as anyone else to fall victim to their own illogical preconceptions or subsequent misperceptions. The result is that biases of many kinds can influence "considered judgment" of the meaning to be extracted from a collection of research reports. For this reason, purely narrative discussions of brain imaging data are coming into disrepute. It is with such narrative interpretations that a multitude of subjective biases are free to run amuck and to be expressed with impunity without the constraining influence of quantitative tests. However persistent informal narrative methods may be, there are countless examples in which narrative prejudgments can distort the inferences to which even the most expert interpreter of complex data may come. In their place, a quantitative evaluation is to be desired if the data are amenable to it. The methods discussed in the following sections all strive to make quantitative and objective that which is often qualitative and subjective.

2.3.2 Simple Voting
In some situations in cognitive neuroscience, it is possible to phrase research questions in a manner amenable to simple voting methods. For example, research might be carried out to answer a question such as "Is there any relation between the behavioral dysfunction known as autism and the size of a particular area of the brain." Brambilla, Hardan, di Nemi,

Perez, Soares, and Barale (2003), for example, compared research reports seeking to answer this specific question and found a discouraging mix of answers. Half of the studies they examined reported a significant correspondence between autism and regional brain size, and half did not. The vote, in this case, was inconclusive.

Similar relatively simple vote counting was used by Sommer, Aleman, Bouma, and Kahn (2004) in their examination of the hypothesis that there existed a bilateral brain difference in language representations between men and women.[2] The result of the voting procedure was that 15 studies reported no difference, whereas 9 found a difference. We should consider this an ambiguous answer to the question of gender differences unless the differences can be explained. In reviewing these experiments, Sommer and her colleagues reported that most of the studies that did show a difference were biased by small sample sizes—the larger the sample size, the more likely it was to find no difference.

The point is that a simple vote provided an ambiguous answer to what initially seemed to be a straightforward question; a more detailed kind of analysis was required to resolve this issue. Voting, in general, but even more specifically when dealing with highly variable data, seems to be an inadequate way to resolve such issues. In this case, identification of a bias (inadequate sample size) was possible; in others, the voting method would leave us with great uncertainty about the answer to what may have initially seemed to be a "simple" question.

2.3.3 Conventional Statistics

Many research questions may demand something beyond a basic binary, yes-no, answer than can be, in principle, resolved by simple votes. Answers to specific questions may have to be provided, such as "What is the correlation between two variables?" Another specific exemplar question is "What is the magnitude of the effect on learning produced by the interval between training and testing?" In traditional experimental psychology, there typically is considerable variability in the answers obtained to such questions. Variability demands that multiple tests be given and the statistical properties of the pool of answers to the question of average effect size and variability be determined. All of us are familiar with means, variances, standard deviations, and the panoply of tests that are used to determine how much credence we can give to data that are pooled by simple statistical averaging.

There is a rich culture of statistical methods for pooling effect sizes in experiments characterized by multiple sources of variance and response variability.

Descriptive statistics is a straightforward method for pooling data, usually within the confines of a simple experiment, but it can also be used also to combine the results of a number of experiments. Averaging of this kind may thus be considered to be a simple form of meta-analysis. A clear example of such a meta-analytical averaging procedure for a purely psychological experiment can be found in the work of Brown and Okun (in press).

There are many difficulties in just averaging, of course. Among them is the obvious fact that effect sizes are not always presented using the same measures. For example, Okun and Brown found that the sample of studies they meta-analyzed reported their findings using three different measures of effect sizes—hazard ratios, odds ratios, and relative risk. Such differences were overcome by interconverting hazard ratios and odds ratios to relative risk scores using methods proposed by Zhang and Yu (1998). Such procedures then permit the investigator to apply standard statistical techniques to the data using a common measure of effect size.

Bangert-Drowns and Rudner (1991) summarized the development of the field using conventional statistics as primitive forms of meta-analyses up to the dawn of the brain imaging era. They suggested the following categories for what we would now consider conventional statistical techniques for pooling data.

- *Classic* or *Glassian meta-analysis*, the standard on which all of the other methods were based. It was characterized by inclusiveness of all possible data, use of multiple findings from each study, and averaging different dependent variables.
- *Study effect meta-analysis*, which used more selective inclusion rules and used only one effect size from each study.
- *Tests of homogeneity*, which used analysis of variance procedures in order to evaluate sampling error.
- *Psychometric meta-analyses*, which combined the best features of the other methods. (Abstracted from Bangert-Drowns and Rudner, 1991)

Not surprisingly, there was vigorous debate among the early meta-analyzers concerning the efficacy of each approach—debate that continues to this day.

2.3.4 Spatial Patterns

The technology of cognitive neuroscience has made demands on the standard statistical approach that often require a completely new set of methods as well as a distinctively different perspective on the part of the researcher. Instead of the single-dimensional effects typical of classical methods, brain images now confront us with two- and three-dimensional patterns in space. The typical fMRI meta-analysis, for example, involves many extended activation peaks scattered throughout the three-dimensional brain. In such situations, the properties of a huge number of pixels, rather than an overall measure of effect size, may have to be considered. Our traditional statistical methods must be expanded to handle this new multidimensional environment.

Consider the situation in which a number of fMRI experiments produce an array of activation peaks in the three-dimensional space of the brain. Each of these places in space may represent a focus of neural activity presumably created by or associated with the psychological task confronting the subject. If life were simple, then all of the experiments that used a similar stimulus or task would produce the same distribution of activity, and we could draw our conclusions from single, well-designed experiments. However, as recent research has compellingly made clear, such simplicity in outcomes is not usual. However well we may control the stimulus, there are many sources of variability that tend to produce an array of what may often appear to be a random distribution of activity. Furthermore, it is frequently the case that each of the several experiments directed at a particular problem produces a different distribution of these peaks.

What is a researcher supposed to do in this case in order to resolve the question of which parts of the brain are activated by a particular cognitive process? One of the simplest things that can be done is simply to plot each of the reported peaks in a three-dimensional space. (For example, see figure 2.2 earlier in this chapter.) Ideally, at least some of the peaks would cluster in a way that would suggest the common impact of the stimulus condition. In actuality, however, few cumulative plots of high-level cognitive processes show such visible clustering. In fact, as shown in the example in figure 2.2, the typical result is a very broad distribution of activation peaks—the breadth of which seems to spread ever more widely rather than clustering in localized regions as increasing amounts of data are accumulated. Several methods have been developed for analyzing distributed patterns of this kind in the search for some kind of "average" or typical response.

Aggregated Gaussian Estimated Sources

One of the earliest methods for meta-analyzing the brain responses to cognitive tasks was based on computer graphic techniques. Chein, Fissell, Jacobs, and Fiez (2002) pooled the data from 30 studies on verbal working memory but restricted their analyses to the brain region traditionally associated with Broca's area. Their novel means of processing was to use a Gaussian distribution space around each activation peak as a weighting function[3] rather than dealing with it only as an isolated point in the Talairach and Tournoux space.

Chein and his colleagues (2002) then simply added all of the Gaussian weighted responses together to give a cumulative and continuous map of the activity induced in this region of the brain by the working memory task. Where peaks were clustered together and mutually strengthened by proximity, the value of the contrast (encoded by them with color) was high, and vice versa. This produced a variable contrast pattern over the designated region. Figure 2.3A is a sample of this distributed activated pattern recorded from a meta-analysis of all 30 studies. The value of any particular point in the pictured space is that obtained from the sum of the Gaussian weighted activations.

The next step was a highly arbitrary one derived from computer vision techniques. Chein and his colleagues (2002) simply applied an arbitrary cutoff level to the cumulative map, the result of which was the emergence of two maxima in this brain region they interpreted as distinguishable (from the surround and from each other) functional regions as shown in figure 2.3B. This process, in the language of computer vision technology, would be called "thresholding."

Chein, Fissell, Jacobs, and Fiez's (2002) special contribution was the introduction of the idea of Gaussian weighted zones around each of the activation peaks that was originally specified simply as a point in the Talairach and Tournoux space. The next step—simply adding the distance-corrected values—provided a means of having nearby peaks mutually support each other more strongly than distant ones, thus introducing proximity as a regularizing process for pooling peaks. In a system with such

Meta-analysis: The Methodology

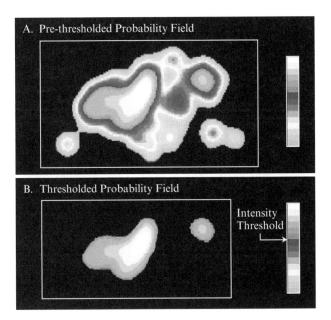

Figure 2.3
(A and B) Demonstration of the AGES technique, a method of computer vision processing. From Chein, Fissell, Jacobs, and Fiez (2002) with the permission of Elsevier Science and Technology Journals.

high positional variability as exhibited in their work, some such means of quantifying "nearness" was essential.

The limitation of such an approach is that there is no objective way to determine where one should set the arbitrary cutoff threshold. Depending on the choice of threshold and the scaling dimensions used, there could be one, two, or even more distinguishable regions. Methods that are more modern at least partially overcome this problem by providing statistical methods to quantify the choice of a threshold.

2.3.5 Statistically More Complex Meta-analysis

The use of qualitative summaries, depictions, graphical approaches, and simple statistics of the results from a number of brain imaging experiments has been supplemented in the last decade by a number of more elaborate quantitative methods. As Turkeltaub, Eden, Jones, and Zeffiro (2002), the creators of the most popular meta-analysis method, have pointed out, there is a fundamental advantage to be gained from quantifying pooled data, thus raising our judgments from the arbitrary and subjective to the objective. They summed up the advantages of formal mathematical meta-analyses as (1) the automatization of the analysis, (2) the quantification of the level of concordance in addition to the location, and (3) the use of significance thresholds, providing statistically defensible conclusions (p. 765).

Nevertheless, the goal of these powerful new methods is the same as that of the simpler methods—to find consistency or concordance among the distribution of responses that were generated by cognitive stimuli in a number of different but presumably related experiments. That is, the goal aspires to determine which peaks are more likely to be common responses to a particular cognitive state and which are irrelevant or idiosyncratic to the single experiment from which they came. The key to making these discriminations is finding where peak activation responses from a number of experiments are clustered. The noise in this system is the position of the widely distributed array of activation peaks both within and without these clusters; the signal is the subtle clustering of a subset of those peaks that is otherwise obscured by the noise. The strategy is much the same as that of the signal detection theory approach using statistical decision-making methods to determine where the clusters are significantly different from the noise. These methods depend on the information that is depicted in the spatial plots shown, for example, in figure 2.2 (Murphy et al., 2003). The basic data for such an analysis are in the form of sets of three-dimensional coordinates, each triplet representing the location of one of the peak activations reported in any of the experiments that were pooled for the meta-analysis.

Thus, the goal of this method is to distinguish a cluster of salient (i.e., concordant) peaks from those that might be considered to be random, spurious, false, or outlier responses. The test for inclusion as a "real" activation is whether it is a part of a cluster of other reported peaks, all of which are supposedly responsive to the same cognitive task. The theory is, in other words, that the more a group of peaks is clustered together, the higher is the probability that they collectively represent a common brain region associated with the common cognitive task. Isolated peaks, on the other hand, were considered to be "noise," "irrelevant," or "random" events that emerged because of differences in the original experimental protocol.

In short, these meta-analytic methods are designed to identify regions of activation that may not be visually apparent in the raw data. They are examples of classic signal-to-noise problems enjoying the advantages and disadvantages of statistical processing. They are necessary because the signal-to-noise ratios may be rather poor in some situations (e.g., see the work of D'Esposito, Deouell, and Gazzaley, 2003, where signal-to-noise ratios as low as 0.10 were observed).

Activation Likelihood Estimates

The pioneering and currently the most often used quantitative meta-analytic procedure of this kind, as noted earlier, was proposed by Turkeltaub et al. (2002); it was first used by them in a meta-analytic study of single-word reading. Their original paper was designed to pick out clusters of peaks that were not apparent in the three-dimensional glass brain as depicted in figure 2.4.

To accomplish this goal they assigned MNI coordinate values to each of 172 peaks from 11 PET studies. However, the peaks were not represented as points. Instead, as Chein, Fissell, Jacobs, and Fiez (2002) did in the aggregated Gaussian estimated sources (AGES) system, they were represented as "localization probability distributions" centered at the MNI point coordinates (p. 769). These probability distributions defined spaces surrounding each point that diminished in their values (i.e., produced lower weights) with increasing distance between peaks. That the probability distribution function fell off with distance acted to diminish the influence of peaks widely separated from each other. This strategy helps to form meaningful clusters (if they exist) because peaks that were close to each other would produce stronger interaction values than those that were more distant.

Each of the probability distribution regions associated with each of the 172 peaks used in the Turkeltaub et al. meta-analysis was then used to calculate the probability that a peak lay within each voxel. Obviously most voxels would have a probability of zero because the voxels were much smaller and more numerous than the spatial extent of the probability distribution surrounding each peak. However, some voxels would have a number of peaks within their space. Every voxel was then assigned a number—the activation likelihood estimate (ALE)—that indicated the joint probability of peaks being present in that voxel.

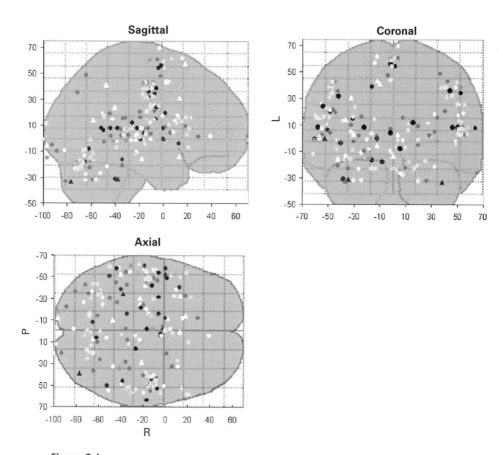

Figure 2.4
A "glass brain" display of the distribution of activations in response to a single-word reading task. From Turkeltaub, Eden, Jones, and Zeffiro (2002) with the permission of Elsevier Science and Technology Journals.

Turkeltaub and his colleagues then compared the ALE values of the experimental data to a Monte Carlo simulation (computed from 1,000 randomly generated sample sets of 172 random peaks). This permitted them to use conventional p values to evaluate the null hypothesis. The null hypothesis in this study was that the distribution of peaks shared the same statistics as the Monte Carlo–generated random background and, thus, there was no significant spatial clustering or concordance of the peaks from the collection of experiments being analyzed. The significant ALE values were then further filtered by establishing a threshold above which they were accepted as significant clusters. These "statistically significant maxima" were then considered to define brain regions that were associated with the single-word-reading cognitive task.

The basic idea was to determine the likelihood of activation for every voxel in their brain space for the 172 peaks and compare it to the random predictions of the Monte Carlo simulations. Using this method, Turkeltaub and his colleagues (2002) identified 11 regional clusters of the brain where the peaks were clustered sufficiently well for them to conclude with a certain degree of statistical confidence that these clusters were associated with single-word reading. These were the regions in which the null hypothesis could be rejected. However, not all clusters were direct indicators of the cognitive activity—single-word reading—under study. Indeed, the strongest clustering activity was to be found in the primary motor cortex, hardly a surprise for a motor task such as oral reading.[4]

As Turkeltaub and his colleagues effectively and collectively argued, the main advantage of this technique is that, in principle, it elevates the evaluation of brain regions from narrative or qualitative interpretations of graphic or tabular information to a quantitative and objective evaluation. However, even they acknowledged that the technique is by no means foolproof. All of the sources of bias, variability, and error that are discussed later in this chapter remain germane in any evaluation of their method. Furthermore, as they also discussed, any correspondence between the ALE-determined clusters and those peaks obtained with a single brain image is likely to be far from perfect.

Because all of the experiments pooled in their meta-analysis were based on PET images, Turkeltaub and his colleagues attempted to cross-validate their method by carrying out an fMRI test of brain activations for a similar single-word-reading task. The correspondence between the two measures—the meta-analyzed PET scans and the single fMRI image—was not exact; they computed a correlation coefficient of .636 between the two measures, suggesting that only 40% of the variance was accounted for. Whether this level of correspondence between the two measures is sufficient to claim equality or even similarity, others must decide.

There is a more fundamental conceptual problem with the ALE approach, however, that transcends how well it works computationally. When one examines the raw data—the locations of the peak values from the pooled studies in figure 2.4—clustering was not visually apparent; the peaks were distributed over much of the brain. To reiterate, the intended goal of the Turkeltaub et al. ALE method and any others like it was to find localized clusters of activity in regions of the brain that can be associated with particular cognitive functions that cannot be detected in plotted data by the human perceptual system. This expectation is being driven by the assumption that cognitive processes are localized, broadly construed, in the brain. The question may be raised whether we are in some illogical and inadvertent way justifying the hypothesis that cognitively significant regions of the brain exist by using a method that ignores the possibility that the brain is actually more distributed and functionally homogeneous. To carry this critique further, could the clusters actually be the result of random fluctuations? Into which category specific clusters will ultimately be thrust depends on the criterion level of significance adopted by the meta-analyst.

The point is that even this highly formal method is not completely devoid of subjectivity. Judgments have to be made of the role of a cluster compared to other isolated peaks that were not combined. Vestiges of subjectivity in this method are inherent in terms of the criteria guiding the choice of the acceptable p levels as well as the arbitrariness of the thresholds defining a peak.

Even more fundamental, however, is that we may be making what I referred to earlier as the "Venn fallacy." In this case, however, it is not a two-dimensional error but a three-dimensional one. Recall that in the Turkeltaub et al. method, the ALE values were defined by the union or common probabilities of the peaks. Thus, only regions of the brain that shared a significant number of peaks were considered cognitively salient in the analysis. This excludes all of the other isolated peaks that appeared in these data. Depending on the exact nature of the Gaussian, Gaussian-like, or spherical distribution around each peak (another arbitrary factor), peaks might become members of clusters. However, those that are not may be as real participants in the cognitive process as the ones that were included in the cluster. The effect of this fallacious three-dimensional Venn fallacy is the same as the two-dimensional one: peaks outside the clusters are treated as noise despite the fact that they might be as psychobiologically significant as those that become a part of a cluster.

Although figure 2.4 (Turkeltaub et al., 2002) shows large numbers of peaks scattered over broad reaches of the brain, the ALE meta-analysis distills these down to a few regional clusters. The argument is that this distillation process works to identify salient localized brain regions associated with particular cognitive processes by distinguishing between "noise" peaks and "signal" peaks. That identification, of course, is based on the premise that there are localized brain functions. The ALE method tends to support this hypothesis by finding localized clusters that may or may not correspond to cognitive activities.

The problem is, as it is with any statistical phenomenon, that even a random distribution of peaks may have "clusters" appearing by the vagaries of chance.[5] The more intense our analysis, the more samples of subjects or experiments, the more likely we are to find significant responses. What all of these statistical methods have in common, therefore, is the unavoidable possibility that there will be false alarms up to the point that the sample size equals the size of the relevant universe. The .05 criterion is a very arbitrary number.

How, then, do we account for all of the other more or less isolated peaks that were not included in the clustering process? Are they simply to be ignored? Are they simply noise to be excluded in the meta-analysis procedure? Answers to questions such as these still bedevil brain imaging studies of cognition. It is possible that the goal of finding clusters may be a manifestation of our propensity to search for localized representations rather than the actual distributed nature of the brain's cognitively relevant responses.

In short, the ALE method, like all of the developments that followed it, is not entirely free of subjectivity or the intrinsic uncertainty of any statistical analysis no matter how impressive the formal structure of the mathematics may seem to be.

Kernel Density Analysis

Wager, Phan, Liberzon, and Taylor (2003) and Wager, Jonides, and Reading (2004) proposed two additional meta-analytical systems for studying brain images—kernel density analysis (KDA) and a modified and improved version they refer to as multilevel KDA (MKDA). Their descriptions of these methods are presented in a more recent article by Wager, Lindquist, and Kaplan (2007).

In many ways, the KDA method is very similar to that developed by Turkeltaub and his colleagues. Both methods defined a region in space around each activation peak as a means of identifying clusters that are considered to be significant indicators of cognitively salient activity. As previously described, Turkeltaub et al. used a three-dimensional Gaussian probability distribution to define this weight of interaction between activation peaks; the nature of the Gaussian distribution determines that this weighting factor is reduced with increasing distance between peaks. Wager and his colleagues, on the other hand, used an equiweighted spherical space in which distance did not matter until one moved outside that spherical space. Both used Monte Carlo calculations to develop a hypothetical random space against which the null hypothesis could be tested (i.e., to determine if there was any difference between the data and the Monte Carlo simulation). The KDA method uses a simple count of the number of peaks in the spherical kernel that exceed the number that would be expected by chance, whereas the ALE method evaluates the probability of a peak (or peaks) for each pixel.

Wager, Lindquist, and Kaplan (2007) pointed out certain other implicit assumptions that minimize the impact of both the ALE and KDA meta-analytical techniques. The three most significant ones are these:

First, the analyst assumes that peak reported coordinates are representative of the activation maps from which they come.

Second, because the procedures lump peak coordinates across studies, study identity is treated as a fixed effect.

[Third] . . . a third assumption of KDA and ALE analyses is that peaks are spatially independent within and across studies under the null hypothesis (p. 154).

The implications of these three implicit assumptions in this kind of meta-analysis are several. For example, Wager and his colleagues (2007) point out that they imply that one study can dominate and thus destroy the validity of the entire meta-analysis.

To overcome the negative effects of these implicit and potentially biasing assumptions, Wager and his colleagues (2007) proposed a modification of the KDA method, which they referred to as the multilevel KDA or MKDA. They suggested that the basic unit of analysis should not be the location of the peaks but rather the experimental study from which the data originally came. Thus, they suggested, "the proportion of studies that activate in a region, rather than the number of peaks, is the test statistic"

(p. 154). In this method, the three-dimensional space of the brain is not populated willy-nilly with all of the peaks regardless of the study from which they originated. Instead, the peaks are first processed to produce a map for each study. Only then are the maps merged and filtered by the application of a threshold to produce a composite activation map of the brain.

Replicator Dynamics

The methods described so far in this section are not the only methods used to seek order in the noisy environment of brain responses. Other mathematical procedures have also been used to meta-analyze brain imaging data. There are two distinguishable goals of all of these kinds of analyses. The first is simply to determine the regions that are activated when a person is challenged with a cognitive task. This is a basic signal-in-noise detection task in which multiple responses are used to determine where high probabilities of significant clusters of responses occur. The second is to link these clusters together into interacting networks determined by coincidence, that is, by showing that there are networks of brain areas that collectively seem to encode a cognitive process by means of their concordant activity. It should be evident that these two goals are not the same.

Neumann, Lohmann, Derrfuss, and von Cramon (2005) proposed a method aimed at carrying out the second of these two tasks—defining systems of interacting regions that are associated with a cognitive task. Their method is based on a matrix analysis system designated as Replicator Dynamics that highlights the co-occurrence of pairs of peak activations in a group of experiments. By means of the matrix, they hoped to be able to determine the strength of the relationships between various regions and thus determine the network with the strongest interconnections; presumably, this is also the most likely network to represent the cognitive process under study.

General Discussion about Meta-analysis Methods

It is quite clear that the formal meta-analysis techniques briefly described here, and others that most certainly will be forthcoming, mainly offer a way to examine the "where" question in the brain. However, it is by no means certain that this is a meaningful question at this macroscopic level of analysis. From the current empirical point of view, it is uncertain that there are specific brain regions assignable to particular cognitive processes. The psychobiological premises underlying the meta-analytic technique may be just another reaction to the implicit and even more basic assumption of localization of systems if not individual sites. In other words, the mathematics of replicator dynamics (and other meta-analytic techniques) may be built on an assumption of the brain localization of cognitive processes that is not justified. Beyond this controversy, we must also reiterate the observation that even if the "where" question could be answered in some empirical sense, it would not answer the question of how cognitive processes are encoded in the brain. Whether answering the macroscopic "where" question is a necessary preliminary to answering the microscopic "how" question is yet to be determined.

The basic empirical fact is that there is still an enormous amount of variability at all levels in the findings obtained with brain imaging studies. It is entirely possible at this stage of the game that the distribution of peaks across and within the brain is not clustered beyond that expected by chance and that our criterion for accepting a cluster as such is much too low. Responses may be much more uniformly distributed across and within the brain (and many of the figures in chapter 3) than is suggested by any of the meta-analytic methods. If this were so, it would represent substantial support for a more holistic or distributed form of representation than the currently dominant localization postulate suggests.

The typical finding (as we repeatedly see in chapter 3) is that meta-analyses do not replicate any better than individual experiments; the results are strongly dependent on the method and the particular selection of data. Clearly, meta-analyses are confounded, conflated, and biased in a number of different ways. This is the topic of the next section.

2.4 Sources of Bias, Error, and Variability in Meta-analyses

Proponents of brain imaging meta-analyses promise to provide a deeper, more quantitative, and more accurate evaluation of the relationship between macroscopic brain regions and cognitive processes. Their approach has already made a fundamental contribution that might change the Zeitgeist in cognitive neuroscience in a profound way. The multiplicity of regional activations associated with cognitive processes in a variety of experiments make it clear that the reported findings do not support the

Meta-analysis: The Methodology 51

idea that localized and functionally specialized regions of the brain exclusively represent particular components of cognitive function. Instead, it has been an almost universal outcome that the more experiments considered the broader and more diverse are the regions of the brain reported to be activated by any given kind of mental activity. Simply put, this suggests that either there is great variability (and, thus, questionable raw results) or the brain responds broadly and diffusely to cognitive stimuli—or both.

Meta-analyses, therefore, provide a compelling counterargument to both old and new versions of the classic postulate that still stalks modern cognitive neuroscience—the association of narrowly localized regions of the brain with specific cognitive processes. The main idea that should be emerging from these empirical findings is that cognitive activity is represented by very broadly distributed brain systems incorporating a multiplicity of interacting regions and centers.[6] It is now virtually impossible for cognitive neuroscience to associate a single isolated (or isolatable) brain region with a single cognitive process. The best empirically based and most plausible current description of results has to be framed in terms of very broad regions of activation. Of course, there remains considerable doubt about what the word "broad" means in this context. Some of the neural systems are reported to involve many widely separated but distinguishable parts of the brain, whereas other reports suggest that essentially all of the cerebral cortex is involved in any or all cognitive processes. Because of this trend, the modern form of phrenological localization that still pervades cognitive neuroscience can no longer be taken seriously.

Breaking the historical conceptual bond to the simple kind of neophrenology that dominated thinking in the twentieth century would be no mean achievement. Nor does it require the most sophisticated meta-analytical technique. Increasingly, during the last decade or so, individual experiments have reported multiple regions of brain activation responding to cognitive tasks. Chapter 3 presents a broad sample of the research showing that many different brain regions are active in any cognitive process.

As sophisticated and complex as the various techniques for meta-analyzing the results of a diverse set of experiments, it must also be remembered that they are not foolproof. Indeed, there are innumerable sources of bias, error, and variability built into any meta-analytic technique that are only partially ameliorated by increased sample size and the supposed objectivity of the analytic method applied. Some of these biases are common to any kind of statistical analysis; others, however, are specific to the brain imaging methods that are the grist for the meta-analytic mills now being discussed.

Several fundamental problems stalk the meta-analytic approach. The first, to which I have repeatedly alluded, is that there may be, in actual empirical fact, no real correlative relationship between the activities of these macroscopic brain regions and stimulating cognitive processes and tasks; all of our data may be based on random processes or those beyond our experimental control. If so, no amount of meta-analytic data pooling will find such a reliable relationship, and all of the present effort may be for naught.

Another problem confounding this seductive idea of data pooling in the pursuit of precision is that it is not always straightforward how one goes about pooling data from experiments that may be heterogeneous in many of their stimulus, analysis, and response aspects. Furthermore, much of the data reported in single studies has already been presummarized, and, thus, much of the original information has been lost. Although it would be far better to carry out a mega-analysis on the original data, this rarely can be done, and, therefore, methods had to be developed for merging the residual information persisting after data have been pooled in the original experiments.

Thus, there are a number of barriers to achieving the goal of increased precision by means of meta-analysis. First, the mind-brain system is extremely complex with many controlled and uncontrolled variables simultaneously influencing the outcome. Second, although stimuli may be easy to control, the cognitive results of a stimulus may diverge greatly from what the experimenter desires; the actual psychological or cognitive activities may be poorly controlled. Third, signal-to-noise ratios are often low, and results are often ambiguous. Given these many sources of variability, it is perfectly understandable why data should be so inconsistent. The dream of those who wish to use meta-analysis is that there is an analogy between simpler experiments in which variability can be overcome by data pooling and those in which much more complicated brain–cognitive relationships are being studied. To accomplish this feat, very specific techniques have been developed to guide the process.

Among the most specific recipes for carrying out a meta-analysis is the one proposed by Rosenthal and DiMatteo (2001). In short, the steps they suggest are these:

1. Define the independent and dependent variables.
2. Collect the studies in a systematic way.
3. Examine the variability among obtained effect sizes with graphs and charts.
4. Combine the effects using several measures of central tendency.
5. Examine the significance level of the indices of central tendency.
6. Evaluate the importance of the obtained effect size. (Abstracted from Rosenthal & DiMatteo, 2001)

Another well-known "recipe" for carrying out meta-analyses was suggested by Cooper and Hedges (1994). Their steps, in brief, included these:

1. Formulate the problem.
2. Gather the relevant literature.
3. Code the studies.
4. Analyze and interpret the data. (Abstracted from Cooper & Hedges, 1994)[7]

Although these directives may seem straightforward, it is immediately obvious that the potential problems associated with each of these steps are deep and profound. Each step taken from the conceptualization of the process to the interpretation of the findings offers opportunity for misunderstanding the nature of the cognitive processes under study, mistakes in processing responses, and, therefore, the introduction of biases, errors, and variability.

In the following section, I consider some of these sources of bias and error plaguing meta-analysis techniques and discuss how they contribute to the production of high levels of variability. I have categorized these sources in terms of the locus of their origins, that is, at the stage of the total experimental and analytic process at which I believe these biases are exerted. Some of these sources of bias have to do with the choices concerning which samples of the available research literature should be included in the meta-analysis. It is at this initial stage that distortions known as selection biases can occur.

A major problem facing researchers when they select a sample of studies for meta-analysis is that not all seemingly relevant experiments are of equal quality. Some of the individual experiments may not be well designed or executed. Furthermore, each of the pooled experiments may actually not share the same scientific goals despite some common vocabulary. It is not clear that all of the studies of "short-term memory" or "emotion," for example, are actually studying the same cognitive processes. The lack of precise definitions of cognitive processes will continuously contribute to the uncertainty that concepts being compared are actually comparable. That which may seem to be the same cognitive process to one investigator need not be the same to another.

It is, therefore, critically necessary that the data being pooled have adequate similarities. Clearly, we want to have all of the dependent variables measured along the same metric. Otherwise, we would, indeed, be pooling and averaging "apples and oranges." As obvious as this fact is, because of the shortage of relevant data, occasionally meta-analytical investigators injudiciously mix experiments that may superficially seem to represent the same cognitive process but actually represent quite different tasks. Indeed, it is possible that, given the multifactorial nature of cognitive processes, no two cognitive neuroscience experiments should ever be considered to be equivalent and suitable for meta-analysis.

Other sources of bias are generated by the intrinsic variability of the sample of subjects. Others are created by technical problems with the MRI systems that are used to produce the results. Still other unavoidable problems are associated with the inevitable quirkiness of the interactions between human beings and mechanical systems.

Not to be underestimated is the fact that the responses of the biological system—the brain itself—are themselves highly variable. Individual brain anatomy (and to an unknown degree, its physiology) may vary from subject to subject and from group to group. Most cognitive neuroscientists agree that here is no a priori reason that every brain must process a stimulus in the same way using the same neural mechanisms at the same brain locations as every other brain.

Furthermore, the statistical techniques used to carry out the meta-analyses are not themselves foolproof. Human intervention in setting criteria, thresholds, and various tests of the significance of a result can profoundly affect what is finally produced by a meta-analysis. Even more fundamental, however, is the fact that statistical analyses are only as good as their assumptions; all too often even the most basic assumptions such as the normality of the response distributions are neither checked nor compensated. Many statisticians would be amused by the cavalier attitude of some neuroscientists in assuming that their data meet the most basic criteria for statistical robustness.

It is important as I undertake this essentially gloomy task of identifying sources of bias and error that I also point out that the general approach of any science is to carry out successively more complex pooling of results in its search for synoptic, explanatory, descriptive theories and general laws. From this pooling is supposed to emerge a progressive pyramiding of knowledge leading to both ever more convergent general principles and more precise answers to the main questions being examined. In the final analysis, the challenge is to determine whether an average signal (as a representative of the general answer to a major question) can be pulled from the noise produced by the following sources of bias, error, and variability in our research strategy.

The degree to which we meet this challenge depends on stable and reliable measurements and a reasonably high level of correlation between the brain and cognitive parameters. Counteracting this goal of reliable data and thus valid scientific conclusions are a large number of sources of error, bias, and variability intrinsic to this complex research environment. Table 2.1 is a partial compilation and categorization of such sources.[8]

Among the most serious sources of bias and inconsistency in this list are the psychological and neurobiological issues—the innate variability due to differences in our understanding of what are cognitive processes and of the complexity of brain mechanisms. Although it is very likely that many of the technical and statistical sources of bias can be overcome in the future, it is not so certain that the psychological and neurobiological sources can be. People are not so easily driven into perfectly standard cognitive states, nor are their brain anatomies and neurophysiologies likely to be uniform enough to permit easy resolution of some of the problems such variability introduces.

Now let us flesh out a few of the more important items in this list. In the following sections, I discuss the major categories and a few of the items that play an especially significant role in the introduction of bias, error, and variability to brain imaging findings.

Table 2.1
Possible sources of bias, error, and unreliability in meta-analytic research

1. Conceptualization Errors
- Wrong level of analysis
- Underestimation of the difficulty of the problem
- Heterogeneity of many aspects of sample of experiments
- High expectations hold for objective methods
- Poor definition of psychological independent variables
- Limits of the subtraction method
- Poor, uncommon, or nonequivalent metrics
- Difficulty in using three-dimensional locations as a dependent variable instead of a simpler measure of one-dimensional effect size
- Confusing transmission codes with cognitive psychoneural equivalents
- Assuming that pooling data will always give you a better (i.e., convergent) result
- Anticipation errors that permit the current consensus to "beg the question"; specifically, the idea that cognitive can be modularized and brain regions localized
- Vested interests (e.g., funding availability and existing laboratory facilities)

2. Selection Biases
- Inadequate search method
- Apples and oranges; diversity of experimental intent
- Nonrefereed or inadequately peer-refereed researches—the "gray literature"
- Publication bias; the "file draw" problem (the disappearance of negative results)
- Required attentional effort
- Sample size differences
- Experiments with multiple tasks mixed with those involving single tasks
- Differences in reported number of peaks
- Underpowered experiments
- Inadequately reported methodology
- Analysis differences
- Quality differences
- Visibility differences (citation rates)
- Mixing PET and fMRI experiments
- Inability to distinguish high-quality from "junk" experiments
- Absence of key information on conditions of experiments

3. Errors in Original Experimental Design
- Nonindependent data (double dipping)
- Lack of permutation analysis
- Loss of single-subject data as a result of pooling and data smoothing
- Low power because of small sample size
- Poor appreciation of differences between sensory and motor, on the one hand, and complex cognitive representations, on the other
- Inadequate control of experimental conditions

4. Cognitive Variability
- Vagueness in instructions
- Lack of stimulus control
- Poorly defined cognitive responses
- Variability in subject's cognitive state and strategies
- Interference from uncontrolled cognitive states
- Subject's affective state
- Restricted subject diversity
- Test-retest intervals

Table 2.1
(continued)

5. Technical Artifacts
- Intended fluctuations in magnetic field strength in fMRI systems
- Uncontrolled variations in field strength
- Subject body, head, and jaw movements
- Uncontrolled blood pressure and other orthostatic factors
- Other uncontrolled autonomic responses
- Magnet acoustic noise
- Equipment signal-to-noise ratios
- Interlaboratory procedural and statistical differences

6. Decision Criteria
- Arbitrary variations in thresholds
- Arbitrary variations in significance criteria
- Overemphasis on peak activations

7. Neurobiological Variability
- Natural variability in brain responses to the same stimulus or task
- Individual differences in brain anatomy
- Individual differences in brain physiology
- Inadequate brain maps
- Inadequate language for localization
- Inadequate localization coordinates
- Fatigue and vigilance
- Variations in physiological noise with magnet strength

8. Statistical and Methodological Errors in Meta-analyses
- Complexity of three-dimensional statistical analyses
- Preanalysis or pooling of experimental data resulting in loss of data (data pooled are data lost)
- Using high criterion thresholds resulting in distributed areas being characterized by the maximum value of the activation peaks
- Using low criterion thresholds resulting in too many or too broad brain areas being considered to be activated
- Mixing variable thresholds or significance criteria
- Nonnormal distributions of data
- Lack of a common metric
- Voxel size differences between experiments
- Different analysis methods
- Attempting to average spatial patterns
- Pooling nonequivalent data
- Artificially elevating type II errors by attempts to minimize type I errors
- Attempting to achieve significance from a pool of nonsignificant studies
- Balance of hits and false alarms
- Ex post facto narrative "just so" stories

2.4.1 Conceptualization Errors

By conceptualization errors, I am referring to that category of a priori "beliefs" that come from some deep and usually covert place in our cognitive schemas. It is there where we keep those assumptions and premises that guide our scientific behavior. This is what we mean by the Zeitgeist—the implicit consensual agreement on issues that we, as a scientific community, take for granted. In actual fact, these beliefs may represent ideas that have been confirmed as well as some that have not yet been confirmed. Here we are on the edge of philosophical speculation—discussion of possibilities and alternatives, but for which we have little empirical support.

One of the most important of these conceptualization errors is the uncritical acceptance that the level at which we *can* measure is the level at which answers are to be found. In the case of the mind-brain relations, our ability to record fMRI or EEG brain signals is widely interpreted as a pathway to solving the great mind-brain problem.[9]

Another conceptualization error is the high expectations that we have for the meta-analysis approach in answering some of our most complex mind-brain problems. The fact that meta-analyses use a formal and "objective" method sometimes clouds the fact that in many cases such methods cannot compete with the incredibly high pattern recognition capability of the human to infer meaningful integrative narrative (i.e., nonmathematical) conclusions from the results of our studies. It is not necessarily the case that these complex statistical methods are going to provide us with anything more than can a critical narrative of the data.

The sheer length of the list of potential sources of bias presented here is testimony to the many obstacles that stand in the way of successful meta-analyses. The assumption that we will converge on a more "precise" answer by pooling a heterogeneous mess of what may well be incompatible findings remains more a hope than an accomplishment.

There are other fundamental conceptualization problems concerning how we think about the technical details of the meta-analytical approach. Beyond the bare-bones general possibility that it just may not work, all meta-analysts are implicitly assuming that the subtraction method can accomplish that which it sets out to do, that is, to distinguish between the salient and irrelevant responses to control and experimental conditions. An argument can be made that the subtraction of the BOLD values obtained in experimental and control conditions, respectively, does not necessarily reflect differences in neural states (see, for example, Sirotin & Das's [2009] work dissociating neural responses from the fMRI BOLD-based response). For several reasons, therefore, it seems plausible, if not likely, that underlying microscopic neuronal network states may differ wildly from each other even though fMRI images indicate no difference. This, of course, remains a matter of considerable contention; however, the point is that a naive acceptance of the differencing procedure may be a major potential conceptual source of bias in meta-analyses.

Other important implicit assumptions and postulates built into the meta-analytic approach usually go unmentioned in the empirical literature. The fact that some portions of the brain (e.g., the sensory primary projection regions and motor pathways) may serve transmission functions while other more central regions represent or encode more elaborate cognitive processes (e.g., decision-making or emotions) raises the possibility that the neural processes and mechanisms used by the two domains may be substantially different.[10] Therefore, approaches, models, theories, and explanations that are effectively used in one domain may not be relevant in the other. Indeed, the extrapolation of ideas and concepts from the relatively simple sensory and motor regions to the "association" cortex may be one of the major sources of bias in the field.

Finally, the most general conceptualization bias of all arises from the assumption that the meta-analysis approach works! Despite the widespread application of this technology in modern psychology and cognitive neuroscience, the validity of this approach has not yet been confirmed. There is still no "proof" that the answers that we get by pooling data are better than the outcome of a single well-designed experiment. One peculiar effect to the contrary, for example, is known as the Simpson paradox (Simpson, 1951), in which the pooling of two significant sets of data leads to an insignificant result. Wager and Smith (2003) offered one explanation for this anomaly. They noted that because of the variability of the position of what may appear as a peak of activation from study to study, there is always the possibility of two highly significant peaks in the individual studies being canceled out (i.e., falling below significance) when the data are cumulated because of slight differences in position. Wager and Smith also argued that particularly noisy control conditions could also contribute to the lowering of acceptable significance levels in the cumulated meta-analysis resulting in increased type I errors—false positives.

In general, as with many other proposed panaceas, the use of meta-analyses raises serious questions concerning their most fundamental assumptions. Indeed, as we see in the later chapters of this book, there is a compelling suggestion that the sources of bias, error, and variability are so great in this kind of cognitive neuroscience that pooling the results of a diverse set of experiments inevitably leads to an increase in variability rather than convergence on a general answer.

In the sections that follow, I delve more deeply into some of the problems involved in using the meta-analytic methods. The continuing question is whether or not these biases are insurmountable or can be accommodated in a way that makes the method work to the advantage of cognitive neuroscience research.

2.4.2 Selection Biases

If the conceptualization biases were disturbing to some, many other cognitive neuroscientists would argue they are only "philosophical" and can be put aside, at least for the time being. However, of much more immediate concern, the basic methodology of a meta-analysis requires that we select a number of related experiments for the analysis. The problem of which criteria will be used to select which experiments are to be included in a meta-analysis thus arises. Despite the fact that there is plenty of practical advice for the selection of research reports for a meta-analysis, much of this advice is useless, and the process is fraught with opportunities to engender errors, biases, and undesirable variability.

Some of these sources of bias and error in the selection process are due to the inappropriate pooling of data that do not fit together for statistical or content reasons. For example, the findings may not be represented along the same dependent variable dimension. In addition, the cognitive processes under examination may not be the same despite vague superficial vocabulary similarities. Beyond the lack of congruity, however, there are some very fundamental errors that mislead or misdirect the most careful meta-analyzer into inadvertently either including irrelevant or excluding relevant reports.

A meta-analysis may be biased from the start by inadequate search methods. Prior to the advent of Google and other search engines, study selection was always a hit-or-miss procedure in which author familiarity, citation rates, or journal prestige could lead to unbalanced selections. Different experimental procedures often led to pooling of experiments that may have initially seemed, but were eventually not, comparable.

The availability of such extraordinary search devices such as Google Scholar and the availability of vast online university libraries help enormously in broadening the search process and bringing together what otherwise might be obscure and, thus, overlooked but relevant reports. However, computerized searches also can lead to bias in the selection process by incorporating items that may have shared only an ambiguous vocabulary. Nevertheless, it is still possible to miss a critical item or include a spurious one. Human judgment is still needed to distinguish between high- and low-quality studies; simple measures such as the number of subjects often are insufficient to determine that decision.

Prior to evaluating the quality of an experiment is the task of finding it. Even the simplest Google search (or those using PsychINFO, MEDLINE, the Web of Science, EBSCO, the Library of Congress, or any other of the increasing list of search engines) will lead to an enormous number of responses. These lists may be redundant and indiscriminate and, thus, all too inclusive. For example, the simple Google search for "fMRI learning" I personally carried out produced about 623,000 results in the latter days of 2010 and each day the list must increase as new research is published.

The list of potential reports to be included can, of course, be pared down by the inserting of more and more qualifiers in the search terms for each list; for example, when I searched for "fMRI learning skill" the count dropped to about 119,000 results—a number that is still too large for the even the most energetic meta-analyzer. However, each such additional search term raises the possibility of the investigator missing truly relevant research as the search becomes more selective. Occasionally, some preselection is effectively carried out by authors of books with encyclopedic or synoptic proclivities and tendencies or even by other previously executed meta-analyses. Another oft-used source of candidate articles are the reference lists of the articles that turned up in the original search or in citations to those articles.

Given the overwhelming number of reports and articles that might be included in a typical meta-analysis, the next task is to extract the truly relevant research from what may be a huge list of irrelevant responses to the search engine's indiscriminate probes. This is no easy task and requires both objective and subjective decision making. To begin, it is necessary to adopt certain filtering criteria to weed out the irrelevant candidates. It is here that judgmental factors must be invoked to establish informal (i.e., qualitative) standards for inclusion or exclusion of each particular candidate research report. Doing so may require the establishment of a coded system in which the various attributes of an article are represented by simple symbols. For example, an "F" might refer to an fMRI-based experiment and a "P" to a PET-based one. A numeral may refer to the number of subjects used. Coding schemes such as this can help to produce some objective and useful criteria for selecting a report for inclusion or exclusion. However, in the final analysis, no matter how simple, complex, or effective the coding scheme may be, human intervention will ultimately be necessary.

Another important feature used for selection is the nature of the task—does it fit well with the overall topic of the meta-analysis? A discerning researcher must also ask whether or not there was a well-defined control condition. The gender and age of the panel of subjects used in the experiment may also be significant. Weeding out duplicates, or even worse, avoiding duplicate uses of the same data in different published articles is also something that is essentially a human task.

Other criteria may have to be used be to distinguish between experiments that have been properly controlled and represent reasonably relevant cognitive tasks, on the one hand, and those that may have only a superficial similarity to the cognitive process under investigation, on the other. Most important of all in the cognitive neuroscience context is that some method must be developed to encode the properties of the imaged response itself. Is the image to be coded simply as the pointlike spatial location of an activation peak or as a distributed region of interest? One might also ask whether it represents a statistical estimate of the significance of the difference between a voxel at one point and some estimate of the average value of all measured voxels.

Thus, there continue to be some systemic problems in selecting which articles are to be included in a meta-analysis. The difficulty of including or not including an article that may have been pulled up in a computer search is nothing compared to the problem of dealing with the cryptic effect of "invisible" research. Among the most disruptive of these invisible sources is what is called "publication bias" or the "file drawer" problem (Rosenthal, 1979). Publication bias is the underrepresentation of studies (usually negative ones) because they were not submitted or not accepted for publication. Not all experiments produce positive results, and yet, few if any experiments are published that report negative results. The result is that an inordinate weight is attributed to published positive results and the ameliorating effects of the unpublished negative findings are excluded from the meta-analysis. These unreported, predominately negative, findings (i.e., those that do not reach the criterion level for statistical significance) are usually filed away and forgotten even when their experimental design may have been impeccable and their negative findings robust. The "file drawers" are filled both by disappointed investigators and by editors who found the absence of a strong enough positive effect to render a paper unacceptable for their journal. Regardless of the source of this kind of publication bias, the effect is the same—the interpretations drawn from the pooled experiments may be wildly off the mark. If we add a large number of negative results to a few positive one, the results may change completely. No one knows how many negative results are out there, but they may represent a massive source of bias.

Publication bias is not a new phenomenon. Dickersin (2005) pointed out that it has been appreciated to be a problem since the seventeenth and eighteenth centuries. As eminent a scientist as Robert Boyle (1627–1691), Dickersin notes, wrote about the problems generated by investigators suppressing some of their unappealing or inconsistent results. Dickersin went on to quote Gould (1987) as asserting that:

> Few people outside science (and not nearly enough people inside) recognize the severe effects of biased reporting. The problem is particularly acute, almost perverse, when scientists construct experiments to test for an expected effect. Confirmations are joyously reported; negative results are usually begrudgingly admitted. But, null results—the failure to find any effect in any direction—are usually viewed as an experiment gone awry. (Dickersin, 2005, p. 16)

The failure to account for or incorporate negative or null studies that have not been published in a meta-analysis is paralleled by a profusion of reports that may have been published in second-tier or unrefereed outlets. This bias results from what is known as the "gray literature." Many profit-making companies have their in-house journals as well as their financial stakes in supporting work in which a positive outcome is desired and, thus, foreordained. Much of this work is of a marginal nature unencumbered by peer review. With the coming of online accessibility has come access to publication outlets that are not constrained by the filters imposed by old-fashioned journals with their boards of reviewers and editors.[11] Thus, some research may rise to a level of visibility that gives them more creditability than they might otherwise deserve. In short, the gray literature gives undue influence to investigations whose conclusions may be more positive than they deserve.

There is the further practical problem of the amount of the reviewer's energy required by the search and selection process. A careful perusal and analysis of potential reports for a meta-analysis are mandatory. In some cases, the stimulus or intent is easy to determine (for example in the search for drug effects, where the dosage and the medical outcome may be relatively precisely measured and stated). In others, for example those exploring cognitive processes, it is not always clear what the investigators were about; words such as "emotion," "thinking," or "learning" may be umbrellas for a host of superficially similar but actually quite different cognitive processes. Similarly, a hypothetical taxonomy of what are considered to be different types of learning (e.g., skill, episodic, rote, conditioning) may obscure the fact that all of these types are actually examples of some more general psychological or neural process.

The situation is further confused by experimental designs in which multiple influential factors are either intentionally or unintentionally confounded. Therefore, it can take a great deal of concentration and careful scrutiny of the stylistic manner in which cognitive neuroscience literature is reported to unearth the actual intent of an experiment. Simple perusal of the titles of candidate articles is likely to be insufficient to determine that two articles are actually examples of the same cognitive process. Similarly, the same words may be used to describe two quite differential experimental goals. The "apples and oranges" criticism—that often investigations are pooled that do not have the necessary correspondence to justify their inclusion in the same meta-analysis—can thus come in both explicit and implicit forms.

2.4.3 Errors in Original Experimental Design

Although the selection process may be biased by the accessibility or visibility of the candidate articles or the difficulty in establishing relevance, there are other subtle factors involved in the original design of the experiments that also tend to introduce bias, errors, and variability into any meta-analysis. In the past 2 or 3 years there have been a number of critical analyses made of the protocols used in carrying out and analyzing the results of cognitive neuroscience experiments that tend to diminish our confidence in the quality of their design.

One of the most subtle but serious experimental design errors is the omission of a random permutation control condition in any experiment. Nowhere is this best illustrated than in the work of Ihnen, Church, Petersen, and Schlaggar's (2009) experiment on gender differences in language. They initially reported that they found some significant gender differences between the fMRI images obtained from two groups of 10 men and 10 women, respectively. However, when they mixed the men and women into two different groups, each of which had five men and five women (which presumably should have washed out the gender differences), they discovered that equally significant, albeit different, regions of the brain were activated. The suggestion is that the gender differences suggested by the images may have been meaningless; that any two groupings of subjects (especially when their numbers are small) may produce what appear to be brain image differences.

Other kinds of questionable experimental design may also lead to biases and errors. A particularly vigorous controversy was engendered in recent years by the work of Vul, Harris, Winkielman, and Pashler (2009) in their review of the literature on what had come to be called "social neuroscience." The field had grown substantially in the last decade with what appeared in retrospect to be some extraordinary findings. Vul and his colleagues noticed what appeared to be unusually high correlations between fMRI images and cognitive reactions in social situations. Indeed, the correlations were so high, given the amount of expected variability in both the fMRI images and the social cognitive measures, that Vul and his colleagues concluded that they must be exaggerated. On an examination of the literature, they found a pervasive design error in 54 reports associating the two kinds of responses. That design error was "double dipping"— using the same sample of voxel scores to both identify the salient peaks and to correlate with the behavioral measures.

This problem has been also been discussed by Kriegeskorte, Simmons, Bellgowan, and Baker (2009) for both brain images and single-cell neurophysiology research. In a recent follow-up, Fiedler (2011) makes a compelling argument that the problem discussed in the Vul et al. paper is far more pervasive throughout science than Vul and his colleagues argued. Indeed, Fiedler asserts that our entire paradigmatic program of research is replete with efforts to enhance effect size by restricting the values of variables so that our results are often "idealized, inflated effects" (p. 163). Much more has to be said about this draconian interpretation of what may be a considerably more serious problem throughout cognitive science than just its application to brain imaging.

It is not known what the impact of these experimental design errors is on the entire field of cognitive neuroscience. However, the substantial inconsistency of the results for what are supposed to be comparable situations suggests that the very validity of many experimental findings may be an issue in the future. It is seems appropriate that we at least consider the possibility that much of brain imaging data being published these days is simply an expression of very complex systems responding to a number of uncontrolled variables in a way that, for practical purposes, may approach randomness.

2.4.4 Subject Variability

Although the conceptual, selection, and design sources of bias in brain imaging work can be controlled to a degree, the major source of variability—that generated by the human subject—is much more difficult to control. The difficulty of controlling the psychological states of experimental subjects remains one of the most important sources of bias, error, and variability throughout the entire cognitive neuroscience enterprise.

The problem is that even the most precisely defined stimulus (for example, a well-metered brief flash of light) can produce vastly different neural and cognitive results depending on variable individual responses and interpretative and situational factors. Even the "simplest" psychophysical responses are not determined solely by the physical dimensions of the stimulus but also by the relationships in time and space between stimuli. Perhaps the most famous of these disassociations of percept and physical stimuli was demonstrated in the work of Land (1977). In this experiment, he showed that the color experience of a patch of light was severely altered by the spatial arrangement of surrounding patches of other colors. Phenomena such as illusions and contrast effects in which the percept does not follow the stimulus are also well known.[12] Furthermore, even the simplest well-controlled sensory study is also influenced by the instructions to the subject. A slightly different formulation of the question or different psychophysical procedures can produce substantially different results.

The problem is further exacerbated in higher-level cognitive studies by the fact that the stimuli and tasks are not anchored to well-defined physical stimuli (as they are in sensory or motor studies) but to interpretations made by the subjects of what their task is supposed to be. It is, therefore, not always certain what cognitive response will be elicited by instructions to think or not think about a particular event. This results in poor stimulus control and eventually noisy, variable, and biased responses from the subject.

The problem of holding a subject's cognitive state constant during the course of an experiment is also always present. Given that human cognitive processes are in a constant state of change (the stream of consciousness described by William James, 1890), it is not only difficult to define the salient part of a stimulus, but also to track what is clearly a moving target—the resulting cognitive activity. In only a few experiments is there any assurance that the cognitive state desired by the experimenter is maintained from trial to trial, from day to day, or, most seriously, from subject to subject. Indeed, in some cases, the results may be paradoxical; for example, any admonition on the part of the experimenter "not" to think about something is almost certain to evoke the forbidden thought. Cognitive state control, beyond that dictated by the simplest kind of sensory stimulation, is most likely an illusion; its absence contributes to the noise and, thus, to the variability and error rate in any experiment that compares cognitive processes and brain images.

Interday replications for a single subject are also confounded by the simple fact of that subject having participated in the experiment previously; the situational environment of the next day cannot be identical to that of the previous day just because there was a previous day. This effect is conceptualized in Bayesian theory in which previous experience or estimates (the prior probabilities) are shown to modify the outcome (the posterior probabilities) of an experiment. In short, people's cognitive responses are not completely controlled by the momentary stimulus; they are also influenced by their previous experiences and other internal influences (such as their affective state) over which the experimenter may have far less control than is usually thought.

I have already spoken of the importance of the relationship between relatively small sample sizes and the power of the experiment. A related issue is the diversity of the subjects in any experiment. In many current studies, the sample population is drawn from a convenient pool of available college students. Indeed, there is an explicit tendency to try to smooth out the variations in the findings from individual subjects by selecting such a homogeneous group. However it is not at all certain that the other factors involved in determining the outcome of an experiment, factors such as smoking or age or educational level, might not introduce further variability into experimental results. Thus, efforts to make the sample of subjects homogeneous have the unintended consequences of making the results less able to be generalized to the entire population. The general problem of unrepresentative, restricted samples (e.g., college sophomores in psychological research) has been discussed in what is now considered to be a classic paper (Sears, 1986). A more recent demonstration of the persistence of this problem can be found in an article by Hogben (2011).

2.4.5 Technical Artifacts

If the uncertainty about the stability of the cognitive response is almost unmanageable, there are many other biasing factors that are much more amenable to further technical improvements. Although different versions of the same model of a MRI system may be designed to be as similar as possible, the magnetic fields of these complex machines can vary from system to system. Magnetic field strength can strongly influence the images that are generated. Similarly, the type of analysis used to reconstruct an image from the radiofrequency signals emanating during the magnetic relaxation period can influence the nature of the reported regional activations.

MRI machines are also famously noisy, and these acoustic artifacts can also unintentionally influence the neural responses by injecting unwanted sensory stimuli. However, it is the behavioral relationship of the subject and the MRI machine that is probably among the greatest source of errors and bias in research of this kind. To provide a good MRI image, subjects must hold themselves stable and unmoving for extended periods of time. Simple muscular movements, for example of the jaw, respiration, or even of the intestines, can also distort an image. So, too, can blood pressure changes, often unavoidable due to the claustrophobia-creating enclosed space of the standard MRI system introduce distortions. Variations in the hardware can also produce poor signal-to-noise relations in which event-related averaging or multiple scanning is required.

A further technical problem is that not all of the parts of the brain are equally easily imaged with fMRI systems. The orbital frontal cortexes, the anterior lobes of the brain over the eyes, are close to the sinuses. The sinuses are filled with air, and the interfaces between air and brain tissue can grossly distort or attenuate the fMRI image (Kringelbach & Rolls, 2004). Early studies with MRI and PET systems were notorious for not being able to produce strong signals from the temporal lobes (Cabeza & Nyberg, 2000).

The positive aspect of these technical sources of bias, error, and variability is that they are all potentially resolvable. New machines, new analytical procedures, and even such simple physiological practices as breath holding can eliminate or minimize many of these sources of bias. This cannot be said for the next category of bias sources.

2.4.6 Decision Criteria

Once the data have been accumulated and are ready for processing, there is another opportunity for bias to influence the conclusions drawn from a research project. This concerns the decisions that have to be made concerning how an investigator goes about interpreting the new data. One important decision concerns the threshold at which a candidate activation will be considered as acceptable as a real one as opposed to a part of the noise. This may be in terms of the contrast level of the image or the probability at which a response will be considered to be significant. Obviously, high thresholds and demanding p levels will tend to restrict the responses to narrow regions of the brain, and low thresholds and modest p levels will produce more broadly dispersed responses. For all practical purposes, such decisions are arbitrary; nevertheless, they can lead to massive differences in the conclusions drawn. Indeed, whether or not an activation peak will be detected is largely a result of an arbitrary decision concerning the threshold.

There is probably no sure method of formalizing the choice of a threshold; the necessity to do so is a direct result of the stochastic properties of the mind-brain problem being studied. Nevertheless, it is critical in this kind of research that it is always kept in mind that the pattern of brain activations is very much a function of key decisions made in the experimental protocol as well as of the brain's neurobiology—the next topic of concern.

2.4.7 Neurobiological Variability

In addition to the relatively uncontrollable psychological states discussed earlier, the neurobiology of the brain introduces many uncertainties that can bias the outcome of even the best-controlled experiment. Among the most notable is that the human brain varies in its gross anatomy from person to person. Although the courses of the major sulci (i.e., the lateral and central fissures) are relatively fixed from person to person, the courses of the minor sulci differ substantially from subject to subject. It has been suggested (Richman, Stewart, Hutchinson, & Caviness, 1975) that the details of the gross anatomy of the cerebral hemispheres are the results of the random buckling of the cortical surface of the brain as it grows within the confines of the rigid and unyielding skull rather than preprogrammed genetic factors. As a result, sample diagrams of the brain may not adequately describe normal intersubject variations in the anatomy of the sulci.

Thus, brain maps are generally inadequate to specify precisely locations on individual brains or the extent of regions of activation as discussed in the introduction to this chapter. Attempts have been made to develop standard maps using key reference points such as the anterior and posterior commissures (Talairach & Tournoux, 1988) and, by this means, locate regions in the brain within a three-dimensional coordinate system. However, although widely used, even these methods are still not perfect in designating brain locations because individual brain anatomy may differ from the standard set by the Talairach and Tournoux coordinates.

At a more microscopic level, it is, of course, not known what neuronal nets are instantiating particular cognitive functions. However, from a purely logical point of view, it does not seem necessary that exactly the same neuronal network processes are always representing or encoding what appear to be common cognitive processes. It is at this level that the greatest mysteries lie. This microscopic neuronal network level is where most cognitive neuroscientists think the mind is encoded in what may be the most complex interaction process of all instead of the macroscopic level to which the fMRI is sensitive. Brain imaging devices are blind to this level of detail; invisible influences biasing the results of an experiment are, therefore, inevitable.

2.4.8 Statistical and Methodological Errors in Meta-analyses

The processing of information from an fMRI system is a computationally intensive and complex task. Not only is there a huge amount of information to be summarized but the raw radiofrequency signals must be processed by elaborate techniques often involving demanding analyses to produce a visual image.

My favorite example of the many steps involved in analyzing fMRI signals is from the work of Newman, Greco, and Lee (2009), a quotation I have used previously (Uttal, 2011). The many computational steps involved in this single experiment remind us of the multiple potential sources for biases and errors to produce highly variable data. Newman, Greco, and Lee describe their methodology in the following words.

> The data were analyzed using statistical parametric mapping (SPM2 from the Wellcome Department of Cognitive Neurology, London). Images were corrected for slice acquisition timing, and resampled to $2 \times 2 \times 2$ mm voxels. Images were subsequently smoothed in the spatial domain with a Gaussian filter of 8 mm at full-width at half maximum. The data were also high-pass filtered with 1/128 Hz cutoff frequency to remove low-frequency signals (e.g., linear drifts). The images were motion-corrected and the motion parameters were incorporated in the design estimation. The EPI data were normalized to the Montreal Neurological Institute (MNI) EPI template. At the individual level, statistical analysis was performed on each participant's data by using the general linear model and Gaussian random field theory as implemented in SPM2. Each event (trial) was convolved with a canonical hemodynamic response function and entered as regressors in the model (Friston et al., 1995). Although there were two phases for each trial (plan and execute) only one regressor that encompassed both phases was used in this analysis. (p. 131)

At each step of their analysis, which may be impeccable in some fundamental technical sense, opportunities for distortion abound. Every time data are "smoothed," "convolved," "normalized," or "filtered," changes are introduced that could lead to different conclusions about what is happening in the brain—if not in this particular experiment, then in many others, not to mention that voxel size, Gaussian filters, and cutoff frequency also represent arbitrary choices.

When meta-analyses are carried out, further opportunities for distortion of the pooled data are introduced. The act of combining or pooling results from different experiments loses considerable amounts of what otherwise might be useful data. Not all experiments use the same measure of activation; simple coordinates or narrative summary descriptions of brain regions introduce considerable uncertainty, as do the variety of analytic methods used in the original experiments. Specific experimental designs attempting to control type I errors may inadvertently lead to an increase in the number of type II errors (Lieberman & Cunningham, 2009). Arbitrary thresholds and questionable procedures to combine spatial patterns may also introduce errors and biases into the data and lead to false interpretations of the phenomena we are seeking to understand. Indeed, the most basic strategic question can also be asked about the entire meta-analysis enterprise—does the effort to achieve significance by pooling insignificant data actually work? The answer to this rhetorical question is an empirical one that lies in the future, a future that is beginning to be clarified by the empirical data discussed in chapter 3.

Potential biases, errors, and or sources of variability of this kind are, of course, not unique to the meta-analyses of cognitive neuroscience data. They are common to any experimental attack on any issue of importance that is characterized by what were from the start heterogeneous and variable data. However, they are of special concern to the meta-analysis approach because of the many conceptual, technical, and methodological steps inherent in the statistical methods that lie between the raw data and the conclusions to be drawn from pools of them. There already have been a number of examples, most notably those of Vul and his colleagues and of Ihnen and his, illustrating that the problem of cryptic biases may be far more pervasive in brain imaging than is generally appreciated.

In sum, the complexity of a psychobiological or cognitive neuroscience experiment is enormous, and the potential for many different kinds of bias and errors is always present. There are, therefore, many reasons why the results forthcoming from our efforts to associate particular brain loci with particular cognitive processes, even when bolstered by the best possible meta-analytic methodology, would be expected to be extremely variable, if not patently inconsistent. My ultimate goal in evaluating and deconstructing the meta-analyses in this book is to determine just how consistent are the results published in the cognitive neuroscience literature in recent years.

2.5 Pros and Cons of Meta-analysis

In the previous sections I surveyed and reviewed a number of the sources of bias, error, and variability that may occur when one conducts a meta-analysis. This list tabulated a number of potential difficulties, many of which were appropriately raised by investigators carrying out studies of this kind. In addition to these self-criticisms (most of which were warnings about potential problems with their own studies rather than with the general meta-analytic approach), there have been other criticisms by critics who looked at the problem from a more general point of view.

From the modern inception of meta-analysis by Smith and Glass (1977), there was considerable criticism of the whole approach; not the least, but perhaps the first, was by Eysenck (1978).[13] Referring to meta-analysis as an "exercise in mega-silliness" (p. 517), Eysenck argued: "The notion that one can distill scientific knowledge from a compilation of studies of poor design, relying on subjective, poorly validated, and certainly unreliable clinical judgments, and dissimilar with respect to all the vital parameters, dies hard" (p. 517).[14]

Eysenck (1994) later published a more reasoned criticism of meta-analysis suggesting that the years had not cooled his pen or his opposition. In a much more specific attack than that provided in the "mega-silliness" article, he pointed out among other problems that:

• The selection rules make the meta-analysis much too inclusive, often weighting the worst studies equally with the best.

• At the same time, having specific criteria meant that some relevant research would be excluded.

• Although linearity of effects is assumed, many effects are nonlinear.

• Effects are probably more often multivariate than univariate. (Abstracted from Eysenck, 1994)

Others joined the battle criticizing the meta-analytic approach shortly after Eysenck's critique including Shapiro (1994) and Feinstein (1995) in the field of epidemiology. Sim and Hlatky (1996), speaking in an editorial of the *British Medical Journal,* argued that the basic idea of pooling small studies to produce higher-power summaries was questionable. Their main argument was that the pooled data from several different experiments often led to results that did not agree with a comparable single, large-scale experiment. They cited a comparison between a meta-analysis carried out on eight small studies and a single large study involving 58,050 patients; the former (Yusuf, Collins, MacMahon, & Peto, 1988) suggested that intravenous injection of magnesium would help 55% of patients to avoid a second heart attack; the latter (Anonymous, 1995) saw no positive effect of this chemical treatment. The disappearance of significant effects between meta-analyzed and mega-analyzed data is known as the Simpson paradox (Simpson, 1951).

Other authors, such as Hunter and Schmidt (1990), although strong proponents of the meta-analytic approach (they have written one of the standard texts in the field), felt the need to enumerate and counter the increasing number of criticisms of meta-analyses. Their counterargument for the "apples and oranges" criticism was especially interesting if not particularly illuminating. They argued that however different the studies may be in terms of the independent variable, they still may have the same dependent variables. This logical argument is analogized to a scale that can weigh "apples and oranges" and produce a joint weight without any problem. They argue: ". . . if there is a meaningful way to associate numbers with apples and oranges, then there will be meaningful ways to compare these numbers. Mean and variance of such numbers are often useful in such comparisons" (p. 517).

To this "logical" argument, they added a methodological one. They suggested that any differences between experiments (which would render them as "apples" and "oranges") could be accounted for by what are called "moderator variables." In other words, studies may be affected by some cryptic variable in a way that makes them seem to fall into separate "apple and orange" categories when actually they are not that different. Hunter and Schmidt counter this apparent problem by noting that: "However, the fact that the studies are "different" in this logical sense does not in any way imply that the studies are different in any way that is relevant to the variables studied in the meta-analysis. The difference may be entirely irrelevant" (p. 518).

The problem with the concept of such a moderator variable is that it may be enormously difficult to identify in the multifactorial type of experiments so common in psychological studies. A very influential moderator variable may be so deeply embedded in the experiments that it remains invisible to even the most efficient analytic procedure.

Meta-analysis: The Methodology

Rosenthal and DiMatteo (2001) provided one of the best-balanced discussions of the pro and cons of meta-analyses for the psychological community. They pointed out that whatever criticisms may be valid for meta-analyses are probably also valid for individual studies. However, from my point of view, their main contribution was identifying other persistent problems with the whole approach. They reemphasized the "file drawer" problem (first highlighted by Rosenthal, 1979) but argued that this was not so much a formal statistical bias as it was ". . . a more serious bias in research sophistication" (p. 66).

On a more formal level, Rosenthal and DiMatteo also noted that there was a potential problem with the nonindependence of effects. For example, all of the articles that come from a single laboratory may have to be considered as interdependent and may have to be combined using rules that are different from those used for data coming from different laboratories. This is an especially insidious bias when the findings from the individual laboratory share data from what are otherwise intended to be independent studies.

Rosenthal and DiMatteo also argued that the "apples and oranges" bias was highly overstated. They joined Hunter and Schmidt (1990) in arguing that there are some situations in which it is useful to combine results from what appear to be quite different experiments. Again, if it is possible to identify moderator variables, then sense may be made of what might initially have seemed to be nonsense.

In a comprehensive review of criticisms of meta-analysis, Borenstein, Hedges, Higgins, and Rothstein (2009) (who are also generally supporters of the meta-analysis approach) summarized many of the criticisms of the approach, each of which they then rebutted. Their list included the following items:

- One number cannot summarize a research field.
- The file drawer problem invalidates meta-analyses.
- Mixing apples and oranges leads to nonsense.
- Garbage in, garbage out.
- Important studies are ignored.
- Meta-analysis can disagree with randomized trials.
- Meta-analyses are performed badly. (p. 378)

Although I have already dealt with some of these problems, there are several that have not been previously mentioned. For example, the first item mentioned by Borenstein and his colleagues is "One number cannot summarize a research field." The issue being raised here is that data are continuingly being lost as more and more data are pooled at all of the levels of analysis that go into a meta-analysis. Repeated distillation of the findings of many experiments may obscure the fact that there were substantial amounts of variation at each level of pooling. In short, most psychological and neurophysiological systems are much more complex than can be captured by a single cumulative measure of effect.

The "garbage in, garbage out" criticism, probably first enunciated by Eysenck (1978) in this context, argues that any attempt to pool studies is unavoidably going to include bad or low-quality research. This "junk" will contaminate the meta-analysis and, even worse, the fact that its insidious effects are being exerted may not be obvious. The antithesis to the inclusion of junk (which depends on the selection and coding rules) is that some high-quality and highly relevant studies may be missed. Proponents of meta-analysis have counterargued that this problem is exactly what the meta-analysis is supposed to overcome by pooling the good and the bad to determine their collective meaning; bad experiments would be characterized as noise and their effects minimized by the cumulative aspects of the meta-analytic procedure. As we see in chapter 3, this is still more of a hope than a reality.

Borenstein and his colleagues (2009) also commented on the general problem of differences between the outcome of a meta-analysis and a single large randomized mega-analysis using the same amount of data—the Simpson paradox. They also alluded to such work as that carried out by LeLorier, Gregoire, Benhaddad, Lapierre, and Derderian (1997) in which large discrepancies were found between meta-analyses and single large-scale randomized "mega-analyzed" experiments. Borenstein et al. argued, in response, that, given the amount of variability in biological systems, discrepancies such as this are to be expected and that under the best of conditions even individual studies would show differences when replicated.

Borenstein and his colleagues concluded their chapter by again pointing out that all of the arguments against meta-analysis also hold true for single studies. The one point with which all must agree is that meta-analyses are often so complicated and the sources of bias so numerous that they are much less likely to be carried out with the same degree of quality control possible with a single study. Given what we discover when we examine the empirical findings produced by meta-analyses and their constituent studies in the next chapter, their propensity for biases and errors obviously will remain a problem for this field of research for many years.

What do these critical comments (and the felt need to counteract them) about meta-analyses tell us in general? First, and perhaps most important, is that the technique is not a cure-all for all of the problems of either conventional psychology or cognitive neuroscience. Second, there is a major loss of what can be very important information (particularly about individual subjects but also about individual studies) during the pooling process. Third, the meta-analytical methods are very susceptible to a wide variety of biases and errors—factors that falsify outcomes, increase variability, and ultimately confuse theoretical interpretations. Fourth, and most disappointing of all, however, is the relative absence of any discussion of the limits of the method. Among these few critical evaluations is the discussion of reliability of fMRI images by Bennett and Miller (2010) discussed in chapter 1. One can hope that their work is a harbinger of other more critical studies yet to come.

The content of this chapter is summarized in box 2.1.

The most important routes to understanding the role that meta-analyses can play in the study of the brain mechanism of cognition are the empirical data itself. In the next chapter, I examine the empirical data in an effort to determine if the process of pooling data leads to a lower degree of variability and a convergence on a common answer to some of the most perplexing cognitive neuroscience questions of our time. In brief preview of chapter 3, they do not seem to work in the way hoped.

Box 2.1
Chapter Summary

> The methods used in meta-analyzing a body of data are fraught with potential sources of error, bias, and unreliability. These arise from a number of different factors, some concerned with the anatomy of the brain, some with the instability of cognitive responses, and some with statistical uncertainties, among many others. It is still a matter of considerable controversy whether or not the ideal of enhancing certainty by progressive pooling of brain image data will work.

3 On the Reliability of Cognitive Neuroscience Data: An Empirical Inquiry

3.1 Introduction

In any science, the most powerful forces in overcoming any doubts we may have of the validity of an observation are the stability, consistency, and reliability of the empirical data. Scientific conclusions are acceptable to the degree that the data on which they are founded are replicable; that is, that independent observers can reproduce the observation, preferably in a public demonstration. The most astounding developments (e.g., cold fusion) have fallen by the wayside when they could not be repeated by independent investigators.

As Carl Sagan was famously reputed to have said: "Extraordinary claims require extraordinary evidence." However, even ordinary claims have to be subject to some minimal tests of reliability and consistency. Yet, in actual practice, rarely is this done. All too often, initial reports, especially those that have some possible applications or relevance to an important aspect of human life, are widely promoted as *the* breakthrough of the month, year, decade, or of all time. Psychology, in particular, with its endless number of phenomena and behaviors, is replete with controversies that arise because what are thought to be comparable experiments often produce incomparable results. In recent years, attention has been directed at the problem of "disappearing effects." Well-established and long-accepted findings that seem to "wear out" as they are tested and retested.[1]

In a problem area as complex as cognitive neuroscience, with its implications for the very nature of what it means to be human and the uncertainty of the meaning of neurophysiological findings, hyperbole can go to extremes. All too often, obscure, but flashy, research is misinterpreted and peddled to an audience in deep need of the solutions to some of humanity's most demanding questions. Sometimes, highly preliminary work is sold to an audience that simply demands the thrill and excitement of what is essentially science fiction. The current context is that even the most outlandish, counterintuitive, and unexpected result can easily make its uncritical way into the popular media. Therefore, it is not just "important" but absolutely essential that any startling, novel cognitive neuroscience result be tested for its reliability before being accepted or published. Obviously, with the pressure for scientific novelty and "discovery," this is not happening. In point of historical fact, intentional replication is a rare bird indeed. Most psychological research, for example, is never exactly replicated.

The purpose of this chapter is to test the implicit assumption that cognitive neuroscience data converge on a common answer by testing the replicability of our findings. I examine the empirical reliability of brain imaging data by comparing examples of research findings at several levels of pooling. Here I ask: Do findings within each level of analysis agree with each other sufficiently well to provide robust support for the cognitive-brain associations now being drawn? This comparison is first carried out at the level of individual subjects, then at the level of individual experiments, and finally at the level of meta-analyses. It is on this pyramid of reliability or replicability that cognitive neuroscience must depend in its quest to distinguish the valid from the invalid and the nonsense from the meaningful. The implicit and explicit expressed hope is that there will be progressively greater agreement as we pool more and more data. Empirical findings suggest that this goal will be elusive.

The degree to which data from individual subjects are consistent will tell us something about how well brain images can be used for diagnosing such cognitive dysfunctions as autism, schizophrenia, and posttraumatic stress disorder. Furthermore, the degree to which experiments agree with each other will give us some insights into how these kinds of data can provide bases for cognitive neuroscience theories of mind-brain relationships.

The most important level of comparison in the present context, however, is the one made between different meta-analyses. It is at this level that the most extensive pooling of data has occurred: a level at which the smoothing process is supposedly most advanced, the data most comprehensive, and at which the convergence on common answers should be the greatest.

Regardless of the methodology and the sample of selected articles to be incorporated into a meta-analysis, it is essential to ask: Do successive applications of these methods actually converge on what are improved answers to fundamental questions?

If the answers are consistent from one comparable set of studies to another, then support is provided for the argument that we are at least finding correlates or biomarkers of our thoughts in the attractive images provided by brain imaging devices. If, on the other hand, results forthcoming from each of these four levels of analysis do not agree, then it may be more difficult to accept that the brain imaging data are actually representing cognitive activity. Indeed, if repeated empirical findings do not compare well, such a result would strongly mitigate even the most widely accepted tenets currently guiding macroscopic neurophysiological measures of cognitive activity. A lack of correspondence would mean that the argument that many of the outcomes forthcoming from a range of investigations varying from the individual study to the meta-analysis may be spurious. Such an outcome would argue strongly, furthermore, that many of the other basic findings in this field are functions of unknown influences beyond our control.

Thus, this chapter is essentially a study of reliability, replicability, and repeatability as studied with a straightforward comparison process. At its most basic roots, it asks: Do comparable studies provide the same answers to the same questions? Thus, it considers issues that cognitive neuroscience shares with other sciences—but to an enhanced degree because the outcomes of mind-brain studies are so variable. This does not mean that there is anything fundamentally different about the physical and neuroscientific sciences. If there is any difference between physics and biology on the one hand, and cognitive neuroscience on the other, it is a matter of complexity. Such problems may not be infinitely complex by formal definition; should they be, they would be considered to be intractable *in principle* rather than *in practice*—a subtle but important distinction that has much to say about the long-term future of our science. Certainly, both the macroscopic and microscopic neural networks of the brain are so complex as to suggest they are at least intractable *in practice* and probably also *in principle*.

Among the most challenging difficulties obstructing straightforward answers to reliability questions is that the output of a cognitive neuroscience study almost always represents a distillation of responses being generated by a multitude of different influential stimulus factors, some of which are known and some of which are unknown. At each stage of the accumulation process, vast amounts of information are thrown away. These "trashed" data, sometimes irrelevant and sometimes of great significance, however, represent a considerable amount of information that might have helped us to determine if we were actually carrying out a reasonable analysis. For example, when data from individual subjects are pooled, all trace of the specific findings from those subjects as individuals is lost. Although the effects of some of this lost information are supposed to propagate throughout the later steps of an analysis, it is usually impossible to reconstruct exactly what were the particularly influential attributes of the stimulus in determining the final outcome. What may have been minor parameters of an individual experimental design may turn out to be major influences on the final outcome of the meta-analysis and yet be irretrievable after a sequence of data pooling.

A related problem in interpreting pooled brain image data is that quite different cognitive states might result from indistinguishably different brain states. Indeed, it is not just possible but very likely that the multifactorial nature of our responses would allow very different cognitive stimuli to produce very similar brain-image response patterns. The reverse is also true. What may superficially appear to be identical stimuli (to the extent that this is possible) may evoke cryptic and/or uncontrolled factors that produce very different neural responses.[2]

In large part, this inability to define accurately the salient aspects of a stimulus may be due to the poor definition of psychological constructs by psychological theoreticians. Unfortunately, psychology has failed cognitive neuroscience by not providing a classification system of well-enough-defined cognitive processes to guarantee that the same cognitive processes are actually being executed when the same stimulus terminology is being used. The current cognitive taxonomy used by psychologists is based on inferences from the results of a long history of psychological experiments; however, it is still one clouded by vague and ambiguous terminology. It is rare outside of the sensory and motor domains for stimuli to be defined specifically enough to guarantee absolute replication of cognitive or neural states. The result is that it is often very difficult to equate two cognitive tasks even though they may be using the same name.

On the other hand, two very different words may be used to describe what is, for all practical purposes, the same cognitive process. The effect of the vagueness of what we mean by such terms as "attention," "learning," "emotion," or any of the countless number of other psychological constructs propagates throughout the entire cognitive neuroscience enterprise. In short, we do not have the analytic skills to parse "decision making" or "memory" or any other cognitive processes from the intricate and interacting complex of processes that constitutes even the simplest mental activity.

A major conceptual issue, therefore, concerns the nature of the relationship between brain maps and the hypothetical constructs of modern psychology. It is not a priori obvious why elements of these ill-defined modules cum hypothetical constructs of psychology should map in any direct way onto the locations or mechanisms of the brain. Our descriptive parsing of cognitive functions, based as they are on inferences from behavior, need not correspond to the way the mechanism of the brain parses these cognitive functions into brain states.

Nor, as we have seen, is it clear that a dynamic system such as the brain will always produce the same responses to even the best-standardized and most stable stimuli from experiment to experiment if not from trial to trial. Given the many potential sources of variability discussed in the previous chapter, the problem is especially exacerbated by the multifactorial nature of mind-brain comparisons.

We are thus left with the basic empirical question, given the limits of what we are able to do in defining and controlling stimuli and measuring responses: Do two or more subjects, experiments, or meta-analyses purporting to search for the neural correlates of the same cognitive processes produce what we can agree is the same answer? Although the problem of inconsistency has been blatant throughout this new brain imaging enterprise, few researchers have attacked the explicit problem by comparing empirical findings from what were hoped to be equivalent studies. Carrying out such comparisons is the purpose of the remainder of this chapter.

3.2 Evaluation of the Empirical Data

My goal now is to evaluate the reliability of the empirical evidence relating brain images and cognitive processes by comparing what are relatively available examples of presumably comparable studies. As noted earlier, this evaluation explores the reliability of brain imaging studies at four different levels culminating in the most complex meta-analyses. In preview summary, a case is made that the brain-imaging approach to the study of cognitive-brain relations is not achieving most of the goals hoped for when the remarkable devices such as the fMRI were first introduced. What our review of the empirical literature does show is a pattern of low reliability at all four levels. Specifically, we have:

1. Moderately variable intraindividual differences

2. Increasingly variable interindividual differences

3. Highly variable experimental summaries in which data from a number of individuals are pooled

4. Maximally variable outcomes from comparable meta-analyses in which data from similar experiments are pooled

3.2.1 Individual Differences

Despite the hope that stable findings would resolve the question of where in the brain cognitive processes are located, the increased use of brain imaging devices such as the fMRI now makes it clear that the contrary is true. The reported location of brain responses to what were considered to be similar cognitive activities do not agree even for one subject! Intrasubject variability exists even when a single subject is repeatedly tested in the same laboratory over a number of days (McGonigle et al., 2000). Only a few investigators (Aguirre, Zarahn, & D'Esposito, 1998, probably being the first), however, considered the questions of intra- and intersubject variability as researchable questions in their own right. Aguirre and his colleagues set the tone for later research on individual differences when they noted that the differences among subjects were significantly greater than those between repeated runs on a single subject.

This work was followed by that of McGonigle et al. (2000), who extensively studied a single subject and showed that there were major differences over a month long period in the responses to a cognitive task even when all other possible aspects of the experimental design were held constant. McGonigle and his colleagues carried out a three-part experiment that used three types of tasks—visual, cognitive, and motor. Their results for the visual and motor tasks were reasonably consistent from day to day. (This is as expected;

I have long argued that sensory and motor activities are relatively stable transmission processes and have excluded them from my general criticism of the representation of higher-order cognitive processes.)

The cognitive task, however, showed much less consistency from day to day as shown in figure 3.1. In this series, the maximally activated brain regions are indicated by dark spots on lateral views of the brain for the 33 sessions over which the experiment was repeated. On the majority of the days that did result in some activation (and not all days did), the activity was mainly restricted to the parietal-occipital region but with distinctly different distributions of the responses for each repeated day. Surprisingly, given the cognitive nature of the task used in this part of the experiment by McGonigle and his colleagues, the frontal lobes were only infrequently activated.

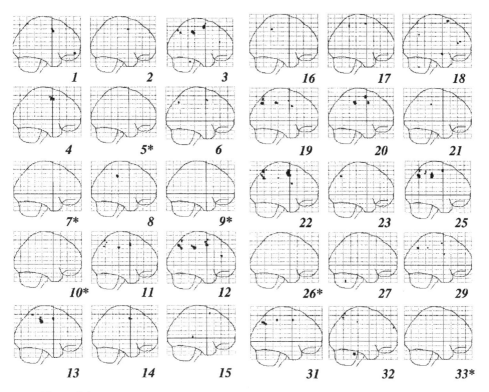

Figure 3.1
Activation peaks from a single subject in a cognitive task taken over 33 sessions. Asterisks indicate no response. From McGonigle et al. (2000) with the permission of Elsevier Science and Technology Journals.

The conceptual impact of this individual's variable[3] responses was heightened by the complete absence of any correlated responses in many of the 30 images McGonigle et al. reported over the course of the experiment. Whether or not a lower threshold or less stringent significance test would have raised or lowered the consistency of these cognitively induced brain images cannot be determined from the data they provided. However, based on these findings, McGonigle and his colleagues emphasized the lack of reliability even within this highly constrained single-subject experimental design. They warned that: "If researchers had access to only a single session from a single subject, erroneous conclusions are a possibility, in that responses to this single session may be claimed to be typical responses for this subject" (p. 708).

The point, of course, is that if individual subjects are different from day to day, what chance will we have of answering the "where" question by pooling the results of a number of subjects? As we see shortly, as great as the changes were from session to session for a single subject, it turned out that that the differences between subjects were even greater.

Among the most active of the few researchers currently studying variability of brain images between subjects are members of Michael Miller's group (e.g. Miller et al., 2002; Miller & Van Horn, 2007; and Miller et al., 2009). They confirmed Aguirre, Zarahn, and D'Esposito's original conclusion that intrasubject variability was less than intersubject variability. It is difficult to quantify this comparison because of the lack of direct comparisons from McGonigle and Miller's laboratories; however, their general conclusions strongly agree.

In general, Miller and colleagues presented a robust and convincing empirical case that differences between subjects were higher than those between repeated trials on the same individual. Miller et al. (2002), for example, compared the fMRI images from a group of nine subjects involved in an episodic memory retrieval task. The results of this experiment showed extensive differences among subjects. Table 3.1 is a matrix of the correlations among six different subjects and between two repetitions on the same subject (the diagonal line). Obviously, the correlations for the two repetitions presented to a single subject are higher than the cross correlations between different subjects. Indeed, the correlations between subjects are very low—less than .30. This matrix strengthened the Miller et al. conclusion that individual subjects produced responses that were more alike on

Table 3.1
Cross (between subjects) and auto- (same subject) correlations showing that differences between subjects are greater than those between repeated trials on the same subject*

	S.C.	K.B.	B.B.	H.G.	C.C.	B.K.
S.C	**.63**	.12	.11	.19	.08	.11
K.B		**.47**	.19	.25	.19	.23
B.B.			**.40**	.29	.25	.25
H.G.				**.50**	.27	.30
C.C.					**.43**	.20
B.K.						**.44**

*Correlations between difference maps at sessions 1 and 2 for the episodic retrieval versus nonretrieval comparison. Boldface numbers indicate the average correlation between a subject in session 1 and the same subject in session 2.
From Miller et al. (2002).

retesting than when compared to other individuals. The low interindividual correlations shown in this table would be rejected by most statisticians as being meaningful.

Figure 3.2, also from Miller et al. (2002), is a more graphic display of intersubject variability. Note the variability of the most significant peaks and how poorly the group average characterizes many of the individual subjects. This figure, like table 3.1, also demonstrates the substantial variability that is observed between subjects. The small circles represent the most significant voxel for each of the nine subjects; some are located in one hemisphere, some in the other, and still others near the midline. One maximally significant response was localized in the frontal region by one subject and one in the occipital region by another. Clearly, there is substantial variability among subjects, so much so that the group average is almost meaningless in terms of characterizing the responses of individual subjects. Equally clear is the conclusion that any average value obtained from pooling the data from all subjects would be a very poor indicator of brain localization for most of the subjects. Only four of the nine subjects were even in the general vicinity of the "average value."

From this general result, Miller and his colleagues joined Aguirre and his colleagues when they concluded that: "Activations produced during retrieval conditions vary significantly from individual to individual, and

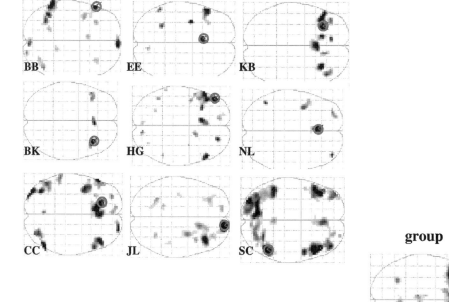

Figure 3.2
A set of nine subjects showing the "most significant" activations and the group average from an episodic memory task.
From Miller et al. (2002) with the permission of MIT Press.

these activations are not adequately accounted for in group analyses" (p. 1209).

Miller and his colleagues, therefore, explicitly argued that the pooling process could lead to grossly distorted interpretations that deviate greatly from the actual biological function of an individual brain. If this conclusion is generally confirmed, the goal of using pooled data to produce some kind of a mythical average response to predict the location of activation sites on an individual brain would become less and less achievable.

In later work, using somewhat different methods, Donovan and Miller (2008) extended this work on individual differences in an episodic encoding and retrieval task. Their results are depicted in figure 3.3.

Even the most cursory examination of these images shows great variation among subjects. Indeed, if one ignores the occipital activations gener-

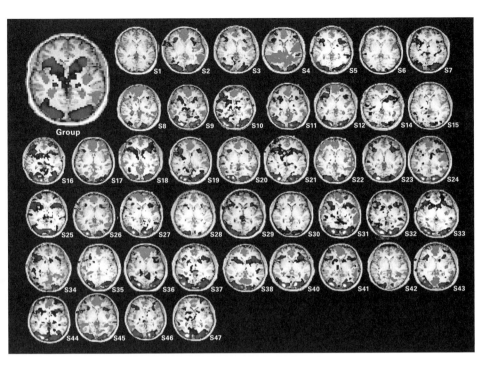

Figure 3.3
Individual differences in episodic encoding and retrieval.
From Donovan and Miller (2008).

ated by the visual task used in this experiment, there is very little in common across this group of subjects as well as between individual subjects and the group analysis shown in the upper left corner.

It is interesting not only to compare the images within this figure with each other but also to compare them with the results of figure 3.2. Despite the fact that the two cognitive tasks were similar, there are vast differences in the representations of episodic memory between the two studies. How much of this is due to the differences in methodology and how much can be attributed to individual differences in the cognitive strategies used to implement the learning task is difficult to specify. However, whatever the specific sources of the obvious differences between these two figures, it is clear that people differ greatly, and findings from one subject do not predict those from another.

The implications of this work are profound. Most important of all, they challenge the idea that, however well or poorly individual subjects may reliably produce brain responses,[4] there is no specific common region of localization across subjects and across time that seems, within the limits of their experiments, to be universally associated with this particular cognitive process. The common theme of a higher level of intersubject than intrasubject variability, therefore, also helps us finally to lay aside the traditional idea that a unique brain location can be found for all subjects for any cognitive process.

The important practical point of both the Miller and McGonigle group's findings is that it is going to be very difficult to identify a pattern of activation responses that can be used as a biomarker for a normal or dysfunctional cognitive process when both intra- and intersubject differences are so large. They also showed that intersubject variability makes it difficult for group data to be used as an indicator of what individual subjects are doing either neurophysiologically or cognitively. An implication of their work is that people must then be examined individually and repeatedly; group averages cannot serve as predictors or signs of individual mind-brain relationships.[5]

Although the observed variability between subjects and sessions are both widespread in terms of the brain-image-based localization of activated areas, there is another level at which individual brain differences are also manifest—brain anatomy. Keller, Highley, Garcia-Finana, Sluming, Rezaie, and Roberts (2007) have shown that the anatomy of the brain is so variable that it is often difficult to localize even the best-demarcated activation from subject to subject. They examined the anatomy of the restricted region known as Broca's area, a region long associated with speech. Broca's area is usually located in the brain regions anatomists call the pars opercularis and pars triangularis. Keller and colleagues, however, found substantial differences in the anatomy of the sulci and gyri in these brain regions from individual to individual. These differences would make it difficult to localize brain regions on subjects even if they were looking at responses to the same stimulus. Anatomical variability, therefore, may also be a contributing factor in explaining many of the individual differences in localization reported by investigators using fMRI responses.

Regardless of the outcome of any debate about the relative stability of individual differences, one general conclusion can come from this discussion—brain image data are very variable. That is, there are both substantial intrasubject and intersubject differences that are obscured

Box 3.1
Empirical Conclusion Number 1

> Individual differences among subjects represent a major problem for studies attempting to relate cognitive process and brain image responses. Although intrasubject variability is less than intersubject variability, the differences both within a subject and among subjects make it impossible to predict from one subject to another or to particularize the response of one subject from the pooled responses of many subjects. If true, this makes any effort to use an individual's brain response diagnostically or to discriminate between different cognitive states implausible if not impossible.

when data from a group of subjects are pooled. The idea of an average or typical "activation region" is probably nonsensical in light of the neurophysiological and neuroanatomical differences among subjects. Researchers must acknowledge that pooling data obscures what may be meaningful differences among people and their brain mechanisms. However, there is an even more negative outcome. That is, by reifying some kinds of "average," we may be abetting and preserving some false ideas concerning the localization of modular cognitive function.

3.2.2 Interstudy Differences

Just as individual differences are obscured by pooling individual data, so, too, are much of the salient data obtained from different studies hidden or destroyed when their results are pooled. In a field of such complexity and with such intrinsically poor signal-to-noise ratios as typify cognitive neuroscience, this is hardly surprising. Nevertheless, the unreliability of what are supposed to be studies of comparable cognitive phenomena appears to be such that there is still considerable doubt about achieving the goal of finding significant, reliable, and useful correlations of brain activity and cognitive processes with brain imaging devices.

As one looks over the field of cognitive neuroscience, particularly that part using brain-imaging devices to generate images of activation areas on the brain, it seems at first glance to be especially prone to unreliable outcomes. Part of this unreliability certainly comes from the difficulty of exactly replicating conditions from one experimental design to another. So, too, do the virtually unlimited options that cognitive neuroscientists have in choosing what to study contribute to this variability. Indeed, it is surprisingly difficult to find truly comparable experiments in this field. Even those tagged with the name of the same cognitive process or those using the same task often differ in subtle ways that prohibit direct comparisons. Furthermore, as we have just seen when we discussed intersubject reliability, the impact of what we might refer to as psychobiologic variability (i.e., a combination of cognitive and neurophysiological variability) also plays a major role in producing inconsistencies and disagreements.

The task of comparing the findings from different studies, therefore, is challenged from the outset by the difficulty of finding truly comparable reports. This is not to say that there are not exemplars of reports that seem at first to be close enough to justify comparison. However, even those that pass a first filter of apparent similarity often turn out to have design features, subtle though they may be, that make direct comparisons questionable. As I have noted many times in the past, slight changes in either cognitive or neuroscientific experimental designs can produce massive changes in results.

One strategy to overcome this variability is the use of a well-defined, standardized task as a second-stage filter in experiments that are to be compared. One example of a moderately well-controlled procedure is exemplified by the Go/No-go test of response inhibition. The Go/No-go test is a frequently used method for exploring executive decision making in which a subject has to decide to respond when one stimulus, the "Go" stimulus, is presented but to withhold a response to an alternative "No-go" stimulus. An advantage of the Go/No-go test is that it is possible to train a subject to respond with different probabilities. By making the "Go" stimulus more frequent than the "No-go" stimulus, the tendency to respond can be enhanced and extra cognitive effort must be made to withhold the response on presentation of the "No-go" stimulus. Because this test is so well defined operationally, it is assumed that any differences in the cognitive and brain responses will be better constrained than in a less well-structured task.

Well-controlled decision tasks like the Go/No-go type generally produce results that suggest frontal lobe activations. However, the "frontal lobe" is a very large structure, and different experiments have reported different activations sites even within this region. (See Simmonds, Pekar, & Mostofsky, 2008, for a discussion of variability in fMRI localization for the Go/No-go task reported by a number of different investigators.) An answer to the question of why this discrepancy still exists was proposed by Mostofsky et al. (2003). They suggested that the neural response differences were mainly due to differences in the details of the Go/No-go task presented to the subject. Such experimental design differences as having multiple "go" stimuli would, they proposed, produce substantial differences in the location of activated areas on the frontal lobes. Different areas of the frontal areas were, according to this explanation, responsive to subtle differences in experimental design, in particular the exact nature of the Go/No-go test.

Although the stimuli are relatively simple in this Go/No-go example, variability is still exacerbated by the uncertain nature of the cognitive strategies being utilized by different subjects. Although something as specific as the Go/No-go task seems relatively straightforward, experiments jointly characterized by this task probably vary considerably in terms of the actual cognitive processes under way as decisions are made—different subjects may have different strategies for solving the cognitive problem posed by the Go/No-go task. Considering all of the potential sources of variability, it not surprising that replication is rare indeed, even when supposedly comparable experimental protocols such as this one are being used.

In another exemplar study emphasizing the response variability observed between experiments, Bush, Shin, Holmes, Rosen, and Vogt (2003) studied the role of the anterior cingulate cortex (ACC) in a number of dysfunctional psychiatric conditions. The general result of these early studies was that fMRI data ". . . produced robust dorsal ACC (dACC) activation in individuals" (p. 60). However, these results could only be produced by averaging data; there were wide differences in the results obtained from individuals. The interpretation of the literature by Bush et al. also generally supported the assertion that response differences between studies are major problems in much modern work. Although Bush and his colleagues went on to claim that they had developed a test (Multi-Source Interference Test) that they believed was able reliably to activate the ACC in individuals, the problem of reliability in other similar studies remains acute. A much better impression of the reliability problem will become evident when we begin to discuss the pooling of data in the next sections.

Other instances of inconsistency in brain region assignments to cognitive responses are abundant throughout the literature. For example, the role of the left inferior frontal cortex in dyslexia is refuted by contradictory reports. Investigators such as Grunling et al. (2004) have reported increased activity in this region, whereas Eden et al. (2004) could not demonstrate this increased activity for dyslexics.

In other experiments, much less well-defined stimulus properties such as "risk" and "ambiguity" are used to distinguish between decision-making tasks. Krain, Wilson, Arbuckle, Castellanos, and Milham (2006), for example, obtained substantial fMRI differences in the responses to these two tasks because it was difficult to define precisely (beyond subtle differences in their respective probabilities) the actual behavioral response differences to the two types of decision-making task.

It is all too easy to review the literature and show that there are many other instances in which what were supposed to be comparable experiments did not agree with each other. However, far more robust arguments for making this assertion can be found in the following discussion in which we more systematically compare the results obtained from a more restricted set of experiments.

I first deal with the simplest kind of data pooling from a group of what are ostensibly comparable studies. This form of data pooling is not a formal or full-blown meta-analysis. Instead, it is a simple comparison of the joint results of a number of related brain imaging experiments by tabulating or plotting them in the same figure, table, or graph. In other words, the data from a number of experiments are simply plotted or listed together. The naive expectation inherent in such a pooling is that as one adds additional experiments, the representation would converge on some kind of an average or high-density cluster that represents the best possible "on the average" estimate of the brain location of the common cognitive process that elicited the results. All that one had to do, it was hoped, was simultaneously to plot the activation pattern results of comparable experiments. This expectation has not been fulfilled. In the place of convergence has come an almost universal dispersion as one adds the data (typically the locations of activation peaks) from additional experiments.

In most noisy or variable measurements of an experimental effect, it is anticipated that as one adds salient data, the values should tend to cluster in a way that leads to an average, central, or convergent value. For example, if one were trying to determine the average height of a population, one would carry out the measurements and then plot them on a graph in the form of a histogram. As additional measurements are entered on this graph, in most situations the cumulative effects would increasingly converge on an improved "average value" of the heights with an ever-smaller proportion of outliers.

Clustering around "average" brain locations, however, is not the typical outcome when comparisons are made of the activation peaks from a number of presumably related brain imaging experiments. In a remarkably high number of such comparisons, using even the best-defined cognitive tasks, the result is that the distribution of peaks increasingly scatters over larger and larger portions of the brain as more and more experimental results are included! Instead of something approximating a normal and monomodal histogram, the distribution of activated brain regions is relatively flat—indicating a near uniform distribution of results.[6] In other words, the common attribute of such figures is that, instead of tending toward a cluster of "average" activations, the more one adds the activation peaks from "comparable" studies, the more of the brain seems to be involved. Whether there are real correlated or "concordant," but invisible, clusters defined by some cryptic property that can be extracted by more formal meta-analysis techniques is discussed in a subsequent section of this chapter. The residual problem is: What do the scattered peaks outside any subtle, statistically defined clusters mean?

First, let us only consider the several ways in which distribution patterns of activation peaks from a number of related experiments can be displayed in simple graphic and tabular summaries. These include:

1. Simple tables listing the anatomical areas of the brain that are populated with the locations of the activation peaks reported by multiple studies.

2. Two-dimensional maps of the brain on which are plotted the reported positions of the activation peaks from a number of experiments.

3. A "glass brain" depiction by means of which one is able to peer through a transparent model of the brain from three different viewpoints and observe the internal distribution of activations as well as the 2-D maps.

4. A projected three-dimensional plot of the Talairach and Tournoux coordinates of the center points of activation peaks such as those shown in figure 2.2.

I now consider some examples of these types of representations and some studies in which they have been used.

Tabular Presentations

The simplest form of plotting the distribution of responses from supposedly comparable experiments is to compile a table of the results. The pioneering example of this type of pooling was the method used by Cabeza and Nyberg (2000). In these tables, they reviewed virtually the full range of experimental psychological phenomena using the PET and fMRI findings available at that time.[7] Table 3.2 shows a part of one of their summaries for three types of working memory—verbal/numeric, object, and spatial. Within each of the three categories, each group of rows represents a group of experiments in which one of these three kinds of working memory was studied. The columns represent the various major lobes of the brain, subdivided into the various Brodmann areas. The various symbol types indicate in which part of the brain the activations occurred. For example, an open circle indicates a "left-lateral" response, and a closed square indicates a "right medial" one.

What is quite clear from table 3.2 is that the responsive Brodmann areas activated by the different experiments are broadly distributed throughout the brain space.[8] For example, verbal/numeric working memory produces activations across all of the regions of the brain, as do object and spatial working memory. Furthermore, all three types of cognitive processes exhibit considerable variability in that different experiments in each of the three groups activate different patterns of neural response. Thus, there is little agreement among experiments, all of which were designed to activate a similar set of brain regions associated with the working memory task.

The major conclusion one must draw from this table (and, as we shall see, all other methods of plotting this kind of data) is that, rather than being localized, the responses from a variety of experiments were distributed over major portions of the brain. Most of the Brodmann regions of the brain are activated by one or another of the experiments, and all three varieties of working memory appear to produce activations across broad swaths of the brain. Therefore, there is no support for the idea that a specific region (or an interconnected system of regions) of the brain encodes each type of working memory—a persistent and probably incorrect premise

Table 3.2

A tabular presentation of data pooled from a variety of PET and FMRI activations invoked by three kinds of working memory

Table 5: Working Memory

Study	Contrast	Frontal											Cingulate					Parietal					Temporal						Occip			Subcort		
		10	9	46	11	47	45	44	6	8	4	32	24	23	31	7	40	39	38	ins	42	22	21	20	mt	37	19	18	17	bg	th	cb
5.1 Verbal/Numeric																																	
Paulesu 93	lett: hold English - Korean							✿ ✿										✿	✿	✿	✿						○					✿
Salmon 96	lett: hold English - Korean						○	○				■					○			○												
Smith 96-1	lett: hold - match						○	○		✣						○	○		○									●	○			
Fiez 96b	wd/nonwd: hold 5 for 40 sec		✿			○	○		✣																								
Jonides 98a	nonwd: hold 5 for 40 sec		✿		●	○	○										●			○								✿	○			
Awh 96	lett: hold - match						○	✿								○	○											●				
Awh 96	lett: 2 back - search						○	✿			■	□				✿	○											✿				
Awh 96	lett: 2 back - rehearse						○		✣							✿												●	■			
Becker 94	wd: 3-wd Rc - rest																		✿ ○ ○	✿	✿				○			● ✿				

Table 3.2
(continued)

Table 5: Working Memory

Study	Contrast	Frontal										Cingulate				Parietal			Temporal								Occip			Subcort		
		10	9	46	11	47	45	44	6	8	4	32	24	23	31	7	40	39	38	ins	42	22	21	20	mt	37	19	18	17	bg	th	cb
Petrides 93b	num: gen random - count	✿		✿									□			□																✿
de Zubicar. 98	lett: read random - alphabetic	○	✿	✿		✿	✿	✿			●	□		■	□		✿	●	●	●	●	●					✿	●				●
Schumach. 96	lett (visual): 3back - search		○			○			✿			✧				✧	✿										✿					
Coull 96	num: update - search	○	○						✧							●	○										✿	□			■	✿
Salmon 96	lett: update - Korean		○														✿	●														
		✿																														
Cohen 97	lett: load factor (0 to 3 back)	●	●					✿	✿	■							✿	●		○												
Braver 97	lett: load (0 to 3 back)	✿	✿					✿	✿	●	○						✿	●														
Smith 96-2	lett: 3 back - search	●	✿			○											✿										✿					
D'Esposito 98	lett: 2 back - search								✿																							
5.2 Object																																
Haxby 95	face: hold 21sec - sm	○				○						□						○				○										
Courtney 97	face: hold 8 sec (regressor)	✿	✿		✿	✿			✿																	✿		✿	✿				

Courtney 96	face: hold 3 - sm	○ ● ● ● ●	❖			❖
Smith 95-1	shape: hold - match	○	❖		○	
Petrides 93a	shape: SOP - mathc	○ ✿	❖	○ ✿ ✿		✿ ✿ ✿
Belger 98-1	shape: hold - ctrl	● ✿	❖	✿		
Elliot 98b	shape/col conj; hold - sm	○	□	○ ● ●	○ ○	
Klingberg 97	patt: altern match - simple		■			
Owen 98	patt: 1 back - sm	● ✿				
McCarthy 96	shape: det rep - sm	○ ✿		✿	○	❖
5.3 Spatial						
Anderson 94	loc: del sacc -fix	● ✿	□ ✿ ❖	○ ● ✿ ○ ●		✿ ✿ ✿ ✿ ✿ ✿	✿ ✿ ●
O'Sullivan 95	loc: del sacc - rest	✿	□ ❖		● ●	
Sweeney 96	loc: del sacc - guided	✿ ❖				
Smith 95-1	loc: hold - match	● ●		●		●

Table 3.2
(continued)

Table 5: Working Memory

Study	Contrast	Frontal										Cingulate				Parietal			Temporal								Occip			Subcort		
		10	9	46	11	47	45	44	6	8	4	32	24	23	31	7	40	39	38	ins	42	22	21	20	mt	37	19	18	17	bg	th	cb
Courtney 96	loc: hold 3 - sm															✿																
Goldberg 96	loc: hold 4 - match		●	○			○		✿							✿											●	✿		○	●	
Owen 96b	loc: seq: hold - sm					●			✿							✿												✿				
Lacquaniti 97	loc point: 2 back - 1 back								□				□	□											■				✿	□		
Owen 98	loc: 2 back - sm		●						✿	●						●	✿														✿	
D'Esposito 98	loc: 2 back - search		●						✿	✿						●	✿			○							✿				✿	
Smith 96-2	loc: 3 back - search 3	○	●						✿							●	✿															
Owen 96b	loc: SOP - sm		●	●	○				✿			■			●	●	✿											✿				
Owen 96a	loc: SOP - sm	○	✿	○	○				✿						✿	✿	✿										○	■	■			
Gold 96	del resp altem - sm	○	○						○	○							○					○						✿	✿			○
Owen 96a	moves: hold 4/5 - point	○	✿	○					○	○		■				✿			●													
Belger 98-1	loc: hold - ctrl	○	●	○								■				✿		✿			●	○					○					
McCarthy 94	*loc: det rep - sm*		●											✧											○							
McCarthy 96	*loc: det rep - sm*		✿																													

After Cabeza and Nyberg (2000).

of cognitive neuroscience until recently.[9] Instead, there appears to be compelling evidence for one or both of the following two conclusions. The first is that brain activity associated with working memory is widely distributed around the brain. The second is that no single area of the brain is uniquely associated with cognitive processes. Thus, it remains problematical whether the variability exhibited in Cabeza and Nyberg's table is due to what are actually neurophysiologically distributed responses of the brain encoding cognitive process or, more disturbingly, simply the reflection of random influences on brain imaging data.

Two-Dimensional Plots

The second means of displaying a pool of raw brain imaging data uses a conventional two-dimensional map of the surface of the brain as a foundation for plotting the location of activation peaks. Cabeza and Nyberg (2000) have also used this method to summarize the reports on working memory shown in table 3.2 in this format, as shown in figure 3.4. This alternative depiction also emphasizes the wide distribution and empirical inconsistency of the data obtained from experiments purporting to be dealing with comparable working memory tasks.

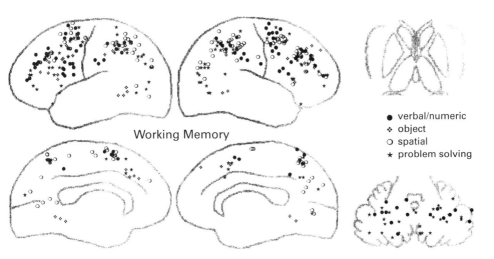

Figure 3.4
A two-dimensional plot of Cabeza and Nyberg's summary of activation areas evoked by working memory tasks.
From Cabeza and Nyberg (2000), with the permission of MIT Press.

This figure is remarkable for depicting the brain responses to working memory tasks as being distributed not only across the entire cerebrum but also within subcortical regions such as the cerebellum. Whatever working memory is, it is not, according to this compilation of data, encoded by the action of a specific localized area of the brain.

Figure 3.5 (from Uttal, 2009b) uses a similar two-dimensional map of the brain as a background for plotting the location of activation peaks reported in 16 experiments studying the effects of deception on the location of brain activations. Although each of the experiments reported only a small number of activation peaks, collectively they produced a distribu-

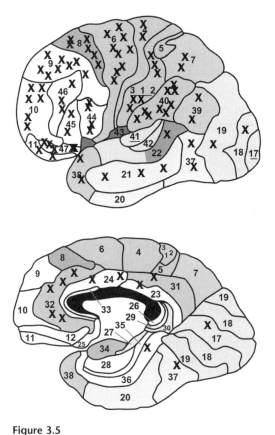

Figure 3.5
The location of activation peaks produced by 16 experiments seeking a biomarker for deception. From Uttal (2009b).

tion pattern over almost the whole surface of the brain. With few exceptions, every Brodmann area of the brain had at least one activation peak in it. Thus, the use of brain images as a substitute for the traditional polygraph (which has its own problems) seems inappropriate. The pictorial pooling of data shown here does not support the idea of any circumscribed region (or, for that matter, any reliable distributed system) as having the potential to be an indicator of deception.

A slightly different perspective has been used by Nee, Wager, and Jonides (2007) to plot the locations of the original activation peaks reported in 47 studies that they subsequently meta-analyzed. Their goal was to locate the concordances associated with interference resolution—a subject's ability to "select information from competing alternatives" (p. 1). Although the geometrical perspective used in figure 3.6 is different, the main point is equally strongly made here—when individually plotted, activations due to what are supposedly common cognitive process are found widely distributed across the brain. In short, there is no evidence for a narrow localization of this relatively well-defined cognitive process. Instead, as one adds experiments, the brainwide distribution of peaks increases. What this says about replicability is obvious.

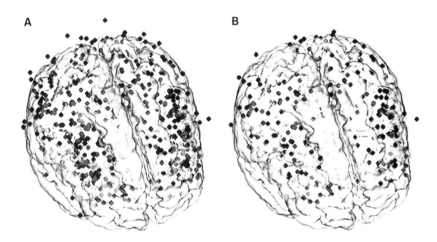

Figure 3.6
Two-dimensional plot of activation peaks from 47 studies used in meta-analysis of interference resolution.
From Nee, Wager, and Jonides (2007) with the permission of the Psychonomics Society.

The Glass Brain

A third means of plotting unpooled data from a number of experiments is by means of a "glass brain." This technique adds a more complete depiction of the activity within the three-dimensional space of the brain by using the Talairach and Tournoux or some similar three-dimensional coordinate system. This is advantageous because the two-dimensional graphic plots just discussed do not take into account that many of the activation peaks may be localized inside the brain rather than on its surface. The glass brain depiction adds the missing dimension in order to produce a more realistic three-dimensional representation. However improved such a depiction many be, it is not easy to put this additional information together perceptually. Each "point of view" must be examined individually.[10]

An example of this technique is shown in figure 2.4 (from chapter 2). In that figure, each plotted point representing a single activation peak (of 172) was reported from a collection of 11 PET studies of a single-word reading task (Turkeltaub et al., 2002). Here, again, the 11 PET studies from which these activation peaks were collected do not seem to agree on any particular three-dimensional localization of the cognitive processes associated with single-word reading; support for a localized region for this cognitive process, therefore, also seems missing.[11]

Another example of a "glass brain" depiction of activation peaks has been published by Kober et al. (2008). They plotted the position of 437 activation peaks from an unusually large number (162) of brain-imaging publications on emotion as shown in figure 3.7. The result of this exercise

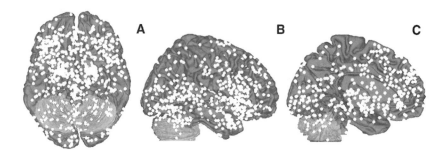

Figure 3.7
Activation peaks from a meta-analysis of emotion.
From Kober et al. (2008) with the permission of Elsevier Science and Technology Journals.

is that virtually the entire brain is participating in emotional activity from at least some of the experiments. The suggestion, once again, is that a cognitive process such as emotion is encoded or represented by activity that is widely distributed throughout the brain.

Even at this simple preanalytical level of examination (in which all of the activation peaks from all of the studies are just individually plotted), it is clear that there is little agreement between and among the reported activations from each of the experiments. Just how random these responses are can be appreciated if we compare the glass brain depictions from two different meta-analytic attempts to analyze the behavioral Stroop test.[12]

The two comparable meta-analyses of Stroop experiments were carried out by Laird, McMillan, Lancaster, Kochunov, Turkeltaub, Pardo, and Fox (2005) and Neumann, Lohmann, Derrfuss, and von Cramon (2005), respectively. The study by Laird et al. plotted the 205 activation peaks from 19 PET and fMRI experiments and produced the results shown in figure 3.8.

Neumann et al. (2005) described the outcome of a similar meta-analysis summarizing 16 experiments on the Stroop test in which 239 activation peaks were recorded. Their results are shown in figure 3.9.

There are both similarities and dissimilarities between the results of these two glass brain plots; however, the general perceptual impression one gets is that the peaks are not distributed in the same way in these two figures.

Once again, the case is strongly made by both of these depictions that almost all of the brain is activated in one or another of the studies included in their respective analyses. If there are any regions in which activation seems sparse, it is the region surrounding the central fissure. Otherwise, there appears to be widespread distribution of the activation peaks reported in both of these studies.

The Stroop procedure is of interest because it is a very well-defined cognitive task that is used here to show that the locations of activations are often inconsistent even when the task is tightly constrained. The result of a comparison of two meta-analyses is that the distribution of the responses differs despite the hope that data pooling would converge on common answers. This is suggested by a preview comparing the results obtained when one carries out the next step in the procedure—the actual meta-analysis. Specifically, Laird et al. identified three areas associated with the Stroop procedure on ALE analysis, and Neumann et al. identified five, as shown in the following lists.

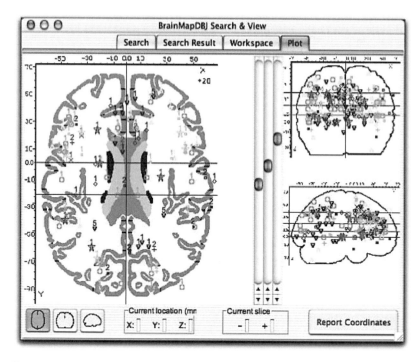

Figure 3.8
A "glass brain" view of a meta-review of 19 studies of the Stroop test.
From Laird et al. (2005) with the permission of Wiley Periodicals.

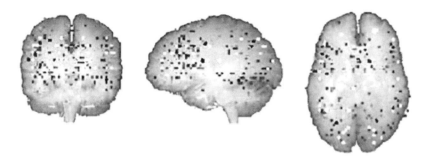

Figure 3.9
A "glass brain" view of a meta-review of 16 studies of the Stroop test.
From Neumann, Lohmann, Derfuss, and Von Cramon (2005) with the permission of Wiley Periodicals.

Laird et al. (2005)
- Anterior cingulate
- Left inferior frontal junction
- Left inferior parietal lobule

Neumann et al. (2005)
- Anterior cingulate (left and right)
- Inferior frontal junction
- Inferior frontal sulcus
- Presupplementary motor area

It is a judgment call about the degree of agreement between these two studies because these two meta-analyses were more complex than exactly comparable tests of the Stroop procedure. Laird et al. were testing both verbal and manual versions using an ALE analysis, whereas Neumann et al. used a different meta-analysis technique—their replicator dynamics method. Nevertheless, with the exception of the anterior cingulate reported by both, the overlap of the areas remains uncertain. The important implication of this is that the answer one gets to meta-analysis questions is dependent on the method. As we see later in this chapter, this lack of agreement between meta-analyses, even when they are using the same method, is a major problem.

Whatever the details of the breadth of the distribution of the reported activation peaks, the most important observation about their locations is that they all illustrate the substantial amount of variability among the various experiments that make up each study. Regardless of any subsequent meta-analysis, which may find some subtle clustering or concordances among these very noisy-looking data, these plots reemphasize three important facts. First, each of the experiments in this pool of data seems to report a different pattern of brain responses. In other words, the data forthcoming from this level of analysis are inconsistent. Second, these activations are distributed over most of the brain. Third, whatever these responses are, they are the product of low signal-to-noise ratios and must be extracted by statistical procedures, procedures that themselves are subject to some question. Should the pattern of meta-analytically defined regions also differ from one analysis to another, the questions would proliferate.

In sum, this section has illustrated an important general fact about comparative studies purporting to associate cognitive processes and the locations of activation peaks on and within the brain. That fact is that, when we are dealing with the raw data (the unprocessed locations of activation peaks), simple graphical plots of the peaks from a number of studies do not lead to an obvious convergence onto a small number of common areas. Far more typical is the observation that although individual studies may report a few localized responses, as one accumulates and plots the reported peaks from an increasing number of experiments, the distribution of those peaks tends to fill up the three-dimensional space of the brain in what appears to be an irregular pattern.

What does this mean? There are several possible interpretations. The first is that the broad distribution of activation peaks encountered in such comparisons emphasizes the lack of replicability of the individual experiments. For reasons both obvious and obscure, these cumulative diagrams reflect an enormous amount of inconsistency among the results from the ensemble of individual experiments. For no other reason than this initial level of inconsistency, the idea that we are converging on answers to the macroscopic "where" questions by means of brain imaging techniques must be carefully reconsidered. Given the apparently uniform distribution of the results in these precursors to full-blown meta-analyses, the argument is strengthened that it is possible that this form of cognitive neuroscience may be based on what are really irregular responses. This suggestion implies that it is possible that there will always be brain image differences between "control" and "experimental" conditions or between any two groups of subjects.[13]

An alternative implication is that these widely distributed responses are real but result from a lack of control over the multiple salient parameters of the stimulus or task. This interpretation suggests that the responses produced in an fMRI experiment, for example, may be influenced by a hidden complex of potential sources of bias capable of hiding the actual cognitive-localization relationships. Cryptic differences in the design of an experiment as well as unknown psychological and neurophysiological differences beyond our measurement, control, and understanding may make superficially similar experiments actually result in quite different, but real, outcomes.

Another interpretation of these results (assuming that they can be validly associated with stimuli and tasks and are not just spurious) is that they are robust evidence for a broadly distributed representation of cognitive process at this macroscopic level of analysis. This would also be an important development because it would argue that there are no particular areas of the brain associated with specific cognitive processes. Rather, most of the brain is involved in whatever cognitive process is under investigation. Explicit in this interpretation is a strong argument against any of the various kinds of localization theories that still permeate the thinking of many contemporary cognitive neuroscientists.

Why, then, one must ask, do the individual experiments so often produce activation in relatively few circumscribed areas? The answer to this question may lie in two non-neurophysiological domains—social psychology and statistics. The social psychological explanation of the sparseness of activation areas reported by individual experiments may reflect the persistence of the belief that the location of a peak of activity is the neural equivalent—the code—for cognitive processes. This is the Zeitgeist; this is the dominant theory; this is what is currently "la mode" among cognitive neuroscientists. As a result, many investigators still tend to direct their quest to the task of finding localized as opposed to distributed responses. It is at this point that the statistical techniques we use can make it possible for these prejudgments to become manifest as "significant" but meaningless results and findings. The implicit assumption that whatever maximum peak or strongest contrast has some special significance can obliterate any effort to consider lesser background activity. The selection of thresholds for acceptance of what we will accept as an activation peak, therefore, can profoundly affect what we observe and what we will eventually conclude from our data.

At this point, I reiterate my conviction that cognitive neuroscientists must acknowledge that there is a nonzero probability that the macroscopic "where" question may be a bad one.

3.2.3 Meta-meta-analyses

It is not immediately obvious that simple cumulative pictorial comparisons of the peak responses from brain imaging experiments, such as those shown in figures 3.4 through 3.9, should or could identify concentrations or concordances of activity in particular brain regions. A compelling argument can be made that the responses are so complex and irregular and with such poor signal-to-noise ratios that these data must be further

Box 3.2
Empirical Conclusion Number 2

> Although individual experiments comparing a cognitive process with the distribution pattern of activated brain sites typically identify a relatively small number of activation peaks, as one adds the responses from many studies, the results are not perceived to converge onto an "average" or localized response pattern. Instead, the brain space tends to fill up with points representing widely distributed activation sites until most regions are filled. This suggests that variability among studies is so large that it is unlikely that any particular pattern from any single experiment is definitive. This result raises serious questions about the validity of each of the individual experiments.

processed to extract any subtle correlations between brain responses and cognitive processes. It is in response to this suggestion that the last decade has seen a proliferation of the much more sophisticated and complex analytical techniques we refer to as formal meta-analyses. The general idea in such techniques is to zero in on a more precise answer to the question of brain localization by combining or pooling the data from as many related experiments as possible in the same way that adding to one's sample size in a simple measurement task can lead to a better estimate of the mean value. The basic idea is simple; the realization of the approach is not so easy.

Meta-analytic techniques are designed to meet this challenge by taking all of the assorted data from what may be a diverse and noisy mixture of experiments and averaging out the "common" or "typical" neural responses. Diversity among the selected experiments is, according to this strategy, counterbalanced by a multiplicity of experimental studies that, when combined, will presumably filter out the reliable from the random.

The typical meta-analysis pools groups of experiments that vary substantially in their design and their results. Furthermore, different meta-analyses typically involve different samples of studies. However, the hope is that, whatever the details of the selected studies are, they are sufficiently germane to the cognitive process under study (e.g., autobiographical memory, decision making, or face recognition) that, when pooled, even the most subtle pattern of clustering can be detected. The basic idea is that our analytic methodology will lead to the detection of common, but narrowly demarcated, brain activation areas that can be associated with particular cognitive processes. This is the conceptual foundation of every meta-analysis—to extract signals from noise by pooling the results of a number of experiments.

The critical empirical question now arising is: Are these "average" or "typical" response patterns elicited by meta-analyses themselves valid and reliable? Do different meta-analyses studying the brain responses to the same cognitive process provide the same answer? The only way we can answer this question is to compare meta-analyses, that is, to carry out a meta-meta-analysis. Should two different meta-analyses aimed at the same cognitive process based on different samples of experimental reports produce essentially the same pattern of brain activations, this would be strong evidence that the hope of correlating macroscopic brain regions and cognitive processes can be fulfilled with this method. According to this scenario, we should be able to overcome the severe challenges posed by low signal-to-noise ratios. A great deal of trust could then be placed in the validity of the meta-analysis technique.[14]

Another important assumption is implicit in evaluating the validity of the meta-analysis technique, namely, that the method used should not influence the outcome. That is, if exactly the same sample of studies is meta-analyzed by two different methods, the results should be the same. If not, any outcome would have to be considered to be method dependent, and no confidence could be expressed in either outcome. Should the outcome turn out to be dependent on the analysis method, this would further raise questions about the hypothesis that there is a meaningful relationship between macroscopic brain locales (or systems of locales) and cognitive processes. Indeed, should different meta-analytic efforts not agree, it would imply that there was something seriously wrong with the entire concept that the findings of brain imaging experiments can be pooled in the same manner as simple statistical averaging to provide an improved estimate of brain activity–cognitive correlations.

Before we compare meta-analyses thought to be attacking the same problem, a few preliminary comments are in order. As discussed in chapter 2, there are a number of meta-analysis techniques that have been proposed. Among the most prominent is the activation likelihood estimates (ALE) method developed by Turkeltaub, Eden, Jones, and Zeffiro (2002). Although other methods have been developed over the years, the ALE method has become the de facto method of choice for most contemporary meta-analyzers. Therefore, both because of the larger availability of meta-analyses using this approach and an ever-present need to focus our analyses, I will concentrate on studies that used the ALE method.

Without repeating the details of the ALE calculation, which are available in chapter 2, I do want to reemphasize what it is that is the ultimate goal of the technique. In short, the goal is to define regions of the brain that consistently exhibit common responses when tasked by the same stimulus. That is, the ALE procedure is designed to determine which regions of the brain reliably respond to a stimulus or task and to distinguish consistent, concordant, or reliable regions from others that are less consistent. As noted, the task is a signal-from-noise detection task; however, it is complicated because we are dealing with three-dimensional spaces filled with inconsistently appearing activation peaks evoked by ill-defined stimuli.

Turkeltaub and his colleagues emphasized the statistical nature of the task. It was not a simple matter of filtering any given attribute or metric as it is in some simple forms of averaging (there is typically no synchronizing stimulus in any of the experiments pooled in a meta-analysis). Instead, the properties of the signal and noise distributions overlap considerably. The discrimination of signal from noise, therefore, depends on the spatial propinquity of a group of neighboring peaks. Specifically, a "signal" (the location of a concordant group of activation peaks) is identified as the center of a sufficiently high-density cluster of activation peaks compared to the low-density dispersion of other "noise" peaks. Thus, it became a problem in estimating probabilities with all of the entire set of attendant difficulties involved in such estimates.

The task was further complicated by the three-dimensional nature of the response space and the small size of the voxels—the unit of spatial extent. The difficulty in carrying out such calculations was further exacerbated by the fact that an activation was usually reported as a point in the Talairach and Tournoux space rather than that which it was in reality—a spatially distributed response pattern. This was partially alleviated by establishing the hypothetical Gaussian space around each Talairach and Tournoux point. How closely such an artifice corresponded to the real pattern of spatial interaction is not known.

The goal of the ALE technique and other meta-analytical methods, therefore, is to define the spatial extent of *consistently* responding regions of the brain as calculated from the respective distributions of related, consistent, or concordant[15] activations, on the one hand, as distinguished from the infrequent and irrelevant (i.e., too distant) activations, on the other. Ideally, it was hoped that this would result in improvement in the definition of the extent of a relatively circumscribed responding region or system of regions of the brain. The extent of these concordant regions is then considered to specify the part or parts of the brain selectively activated by a particular cognitive stimulus or task. It may therefore be correlated with or even (erroneously, in my opinion) identified as the locus of the neural equivalent of the cognitive process under examination.

An important step in the logic of this discussion is that we have to acknowledge that the experiment selection process is working behind the scenes in a deleterious way that cannot be avoided. Two meta-analyses carried out on what is intended to be the same cognitive process may consist of quite different samples of individual studies. However, an implicit idea behind these kinds of meta-analyses is that the sequence of pooling and averaging (from the individual subject through the sample of experiments to the pooling of the mixed bag of activation peaks) should eventually result in an outcome that represents a valid neurobiological correlate of the cognitive process under investigation. It is hoped, rather than proven, that this is true no matter how diverse the subjects, the experiments, and the distribution of the activation peaks were. In other words, the sequential pooling should converge on an "average value" of the salient brain regions in which all of the variability of individual subjects, experiments, and peaks washes out in the sheer amount of data that has been successively pooled. The robustness of this argument depends on the degree to which meta-analyses agree.

The critical next empirical step in testing the validity as well as the logic of this discussion, therefore, is to search for consistency, not at the level of the discrete activation peaks but, rather, by comparing meta-analyses themselves. In other words, we are asking whether or not comparable meta-analyses agree with each other. Do they converge on a common answer? This, of course, is a very robust test of the whole enterprise of seeking answers to the localization question by progressive pooling of data from the several levels of combination. A lack of interexperimental consistency is the prime force leading to the meta-analysis technique. However, should the meta-analyses not agree, it would also suggest that the meta-analytic techniques themselves might be forcing regionalization in situations in which it is not justified. By "forcing," I mean that an answer to the localization question is being imposed by the method rather than by the data. Because these techniques are designed to find consistency, they can do so even in situations in which that consistency might not exist!

My point is that using a tool that is designed to carry out a particular operation can in some cases impose its own properties onto an otherwise neutral situation. The question of the consistency of different meta-analyses of a reasonably well-defined cognitive process, therefore, rises to a particularly high level of importance. How we carry out these comparisons is critical in understanding the outcome of such an examination. Comparing meta-analyses is not a simple process, and there are many opportunities for differences to emerge. Therefore, we should not expect exact replications. The best we can do is to simplify the comparison process as much as possible. The easiest way to do this is to require that all comparisons use the same methodology. All of the meta-analytic studies considered in the following discussion, therefore, were, as noted earlier, selected because they used the ALE technique. There are, however, many other factors that can influence this comparison that could not be controlled; it is impractical in retrospective reviews of the present kind, for example, to demand that they all use the same sample of experiments in their pool or to demand the same panel of subjects.

To put it as simply as possible, my strategy in this section is to find meta-analyses that are aimed at the same topic of cognitive activity and to compare the brain regions that each identifies as being associated with that cognitive process. To the extent there is agreement between the two, I classify them as consistent; to the extent that different regions are identified in each meta-analysis, I classify them as inconsistent. As we subsequently see, inconsistency is the dominant outcome of these comparisons.

The task of comparing two meta-analyses for consistency, as noted, is not an easy one. Laird, McMillan, et al. (2005) proposed one way to do so. Their method requires differencing the basic ALE data (i.e., the probabilities of a peak having occurred in each of the pixels of the image). It requires that the number and locations of the peaks being used in the analysis be known.

Laird et al. argue that it is desirable to pull these values as different subsets from a single meta-analysis—their comparison is, thus, being made on a split-half basis in which all of the data comes from a single meta-analysis thus enhancing the degree of control in the comparison. It is on this detailed basis of the basic data itself that a comparison is made. (For details, see page 158 of Laird, Fox, et al., 2005.) Even then, the method is problematic; according to Laird and her colleagues:

Caution should be exercised when carrying out formal comparisons of ALE meta-analyses when the groups are disparate in total number of foci. In these cases, it is impossible to say with any certainty whether the difference maps reflect activation difference across groups of studies or simply show the effect of one group having a greater number of coordinates. (p. 163)

Unfortunately, the detailed data required for this kind of comparison are not generally available when one compares the results of completely independent meta-analyses.

Split-half designs to test for consistently are not ideal, however. The same method is being used, and it is not always clear where the split was made. Was it between two parts of the same data or was it formed by dividing up the experiments into two parts? In any case, the same method is being used despite the assumption that progressive pooling should make the method inconsequential. A more robust test would involve two separate analyses using two different methods.

How, then, did I carry out practical details of the meta-meta-analysis comparing individual meta-analyses in this section? The actual selection of comparable meta-analyses was quickly determined by the relatively sparse number available for a given cognitive process. At the time this book was written it was extremely difficult to find comparable meta-analytic studies. Therefore, with one exception, all of my meta-meta-analyses are carried out on two meta-analyses. That exception is an abundance of such studies for working memory, for which there have already been a considerable number of meta-analyses. I am sure that this situation will be alleviated as the years go by, but any problems with my samples are due to the limited availability of such studies rather than to any intentional selective bias.

My selection of meta-analyses was even further constrained by the manner in which the data findings were reported. As noted earlier, there were three main ways in which spatial information could be presented: (1)

narrative descriptions (e.g., using words such as "superior temporal cortex"); (2) the numerical Talairach-Tournoux coordinates (e.g., $x, y, z = 34, 18, 2$—in this example, the right insula); and (3) Brodmann areas (e.g., BA 21, which is part of the right superior temporal gyrus).

I chose mainly to use the older traditional Brodmann areas as the core measure for most of the meta-meta-analyses presented in this chapter. This provided a basis for a relatively simple preliminary comparison of the degree of congruence between comparable meta-analyses. For a few of these meta-analyses, lists of non-Brodmann areas were available. In a few supplementary reports only the narrative descriptions of the salient activation regions were available. I present these with the appreciation that these narrative lists provide an even coarser means of comparison than tabulations of the Brodmann areas.

My strategy is to develop a simple quantitative score of comparability. I first count the number of areas in which either of two meta-analyses indicates that an area of interest has been established. I then count the areas in which both of the two meta-analyses agree that a concordant response was present. Next, I compute a simple reliability score by dividing the number of areas in which *both* studies report activations by the total number of areas activated by either study.

Obviously, the simple numeric reliability score I use in this chapter is a very, very rough means of comparing meta-analyses. It is nowhere near the statistical sophistication of such methods for scoring reliability as the interclass correlation coefficient (ICC) mentioned in chapter 1. Nor can it be because we do not have numerical values for the data by means of which we can compute the "within" and the "between" variances. However, as a first approximation, this rough measure seems useful in evaluating the hypothesis that meta-analyzing data will converge on a reliable and consistent answer to the question of what parts of the brain mediate specific cognitive functions.

Of course, this is not a perfect strategy. Among other deficiencies, the BAs used as the main measure are not regularly shaped, and a response area may fall into two quite different BAs on the brain even though they are actually relatively close to each other. Alternatively, two activation peaks may be quite far apart and be attached to a single BA because of the elongated extent of the BA. Similarly, some investigators report that certain subcortical or brainstem (non-Brodmann) structures such as the thalamus or cerebellum are activated in their meta-analysis; in only some instances have I been able to compare these noncortical areas. Nor for that matter have I made any attempt to account for laterality effects; regardless of whether a response was reported to be activated in the left brain, the right brain, or bilaterally, all have been combined into a single BA designation.

All of these limitations are acknowledged at the beginning of this discussion. Nevertheless, I claim that however limited and constrained these comparisons may be, they represent a useful preliminary rough estimate of the reliability of meta-analytic studies in modern cognitive neuroscience. After all, if the two studies do not agree to at least a first approximation, there is little value to going into an elaborate statistical evaluation.

Given the high hopes and logic of the concept of progressive pooling of data, as we now see, it will disappoint many who would have expected that methodological diversity in cognitive neuroscience could be overcome by increasing the virtual sample size.

Cognitive processes to be considered include:

- Autobiographical memory
- Emotional face processing
- Single-word reading
- N-Back working memory
- Emotions

Autobiographical Memory
The first meta-meta-analysis I present comes from the field of autobiographical memory. I was able to locate two meta-analyses that appeared to deal with this cognitive process (Svoboda, McKinnon, & Levine, 2006; Spreng, Mar, & Kim, 2008). Table 3.3 depicts the results of this meta-meta-analysis.

In this comparison of 35 locations identified by either meta-analysis, 19 were populated by both, for a computed reliability score of 54%—a modest level of commonality. However, there are mitigating factors that suggest that even this modest reliability score may be misleading. The Svoboda et al. (2006) results report activation of many more parts of the brain than does the Spreng et al. (2008) meta-analysis. This result might

Table 3.3
A comparison of two meta-analyses of the brain activations associated with autobiographical memory

BA	Meta A	Meta B
4		X
6	X	X
7	X	X
8		X
9		X
10	X	
11		X
13		X
17		X
18		X
19	X	X
20	X	X
21	X	X
22	X	X
23	X	X
24	X	
25		X
27		X
28	X	X
29		X
30		X
31	X	X
32	X	X
34		X
35	X	X
36	X	X
37		X
38	X	X
39	X	X
40		X
41		X
44		X
45	X	X
46	X	X
47	X	X
Total BA	35	
Common BA	18	
%	51	

Meta A shows data from Spreng, Mar, and Kim (2008), and Meta B shows data from Svoboda, McKinnon, and Levine (2006).

be due to a lower threshold used by Svoboda's group as a criterion for accepting activations. Indeed, so much of the brain was activated, according to Svoboda, that it suggests that there was no specificity; instead, this meta-analysis alone argues for a general, widely distributed response of virtually the entire brain. Any commonalities with the Spreng BAs under these conditions would be meaningless. Whatever the sources of this discrepancy, it is clear that these two meta-analyses provide very different answers to the question: Which areas of the brain are activated in processing autobiographical memory?

Some additional evidence for the inconsistency of these meta-analyses comes from the non-Brodmann areas mentioned by each group of investigators. Svoboda et al. (2006) mentioned the amygdala, basal ganglia, thalamus, brainstem, and cerebellum in their review as being activated. On the other hand, Spreng et al. (2008) only mention the thalamus and the amygdala in their summary chart. Obviously, these two meta-analyses do not agree to any level that we might call consistent.

Emotional Face Perception

My next example of a meta-meta-analysis compares the findings from two meta-analyses (Fusar-Poli et al., 2009; Li, Chan, McAlonan, & Gong, 2009),[16] each of which dealt with the perception of emotional faces. Their respective findings are plotted in table 3.4.

In this figure, we see that the results of two meta-analyses, supposedly of the same cognitive process—the perception of emotional faces—most assuredly do not agree. Despite the fact that each of these meta-analyses is pooling the results of a number of individual experiments dealing with what their authors presume to be cognate studies, the results of their efforts diverge greatly. Specifically, although the two experiments jointly identify 12 BAs in which activations were reported, only 2 of these (BA 19 and 37) were common areas generated by both meta-analyses, a reliability score of only 14%. Even then, the discrepancy between the two may be seriously underestimated because these meta-analyses dealt with visual stimuli, and both of the common areas were visual areas.

As far as the non-Brodmann areas are concerned, Li and his colleagues mention the parahippocampal gyrus, the lentiform nucleus, and the amygdala, whereas Fusar-Poli and his colleagues mention the parahippocampal gyrus, the lentiform nucleus, and the cerebellum.

Table 3.4
A comparison of two meta-analyses of the brain activations associated with emotional face processing

BA		Meta A	Meta B
6		X	
8		X	
13		X	
17		X	
18		X	
19		X	X
20		X	
21		X	
22		X	
23		X	
30		X	
37		X	X
40		X	
47		X	
Total BA	14		
Common BA	2		
%	14		

Meta A is data from Fusar-Poli et al. (2009), and Meta B shows data from Li, Chan, McAlonan, and Gong (2009).

Single-Word Reading

Next, I compare two meta-analyses of single-word reading as a test of their reliability. The first was presented in the groundbreaking article by Turkeltaub, Eden, Jones, and Zeffiro (2002) in which the ALE meta-analysis technique was first introduced. The other was a more recent study in which stuttering was the target of interest but for which normal control, single-word reading data were also presented (Brown, Ingham, Ingham, Laird, & Fox, 2005).

The results of this study are presented in table 3.5. Once again, we see that there are major differences between the two meta-analyses. Of the 11 areas that were identified by the ALE analyses used in both studies, only 5 areas were common to both. Once again, this indicates a reliability score of only 45% of the identified areas emerging from the extensive data pooling that characterizes these kinds of meta-analyses.

However, once again, the situation may be much worse than even this preliminary comparison suggests. If we examine the common brain regions

Table 3.5
A comparison of two meta-analyses of the brain activations associated with single-word reading

BA		Meta A	Meta B
4		X	X
6		X	X
10			X
17			X
18			X
19		X	X
21		X	X
22		X	X
37		X	
42			X
43			X
Total BA	11		
Common BA	5		
%	45		

Meta A shows data from Turkeltaub, Eden, Jones, and Zeffiro (2002), and Meta B shows data from Brown, Ingham, Ingham, Laird, and Fox (2005).

in greater detail, we see that, as in our previous comparison, some of them may be associated with activities that are secondary to the cognitive aspects of the task. For example, Brodmann areas 4 and 6 are motor areas that would presumably be activated by any kind of motor response —such as those involved in speaking. Thus, although they may be a necessary part of the reading process, they are really incidental to the cognitive process under study. On the other hand, BAs 19, 21, 22 are regions of the brain well known to be associated with visual information processing and thus may be nonspecific with regard to this single-word reading task.

One interpretation of this pattern of responses, therefore, is that the authors have simply identified regions responding to the sensory inputs and motor outputs involved in single-word reading. Once we ignore these obvious common regions, the remaining question is: Why do the additional regions (mainly those reported by Brown and his colleagues—BAs 10, 42, and 43) not show up in the Turkeltaub et al. results? One obvious answer to this question is that the component studies of the two meta-analyses and the meta-analyses themselves used different thresholds of acceptance. However, it is also possible that the differences are due to the quasi-random nature of the brain images and that they are not really correlated to this or any other cognitive process. Whatever the reason, beyond the general motor and sensory aspects of the task, these two meta-analyses do not provide evidence of replication for what was a simple well-controlled cognitive task.

With regard to the non-Brodmann areas that may be involved in single-word reading, Brown and his colleagues mention only the cerebellum, whereas Turkeltaub and his colleagues listed in addition to the cerebellum, the thalamus, and a general region of the superior temporal gyrus without Brodmann designation.[17]

n-*Back Working Memory*

Our next meta-meta-analysis is a two-way comparison of meta-analyses dealing with working memory as particularized by a specific experimental task—the *n*-back test. This is a particularly well-defined cognitive task for which there is a rich background of purely psychological studies. The two meta-analyses I have chosen to compare in this case were published by Owen, McMillan, Laird, and Bullmore (2005) and Glahn et al. (2005),[18] respectively. Although, there were differences in the intent of each report, in the sample of experiments being meta-analyzed, as well as in the methodology used in each of the constituent studies, both shared the common goal of trying to determine which brain regions were activated during the cognitive process using the *n*-back method.

The BAs reported in each of the two meta-analyses are depicted in table 3.6. Despite the relatively well-controlled nature of the working memory task examined here, of the 12 areas of activation that emerged from the two meta-analyses, there is an overlap or correspondence of only five of the BAs from the two meta-analyses.

To summarize the results for these two *n*-back meta-analyses, I computed the rough reliability score—the number of BAs in which both meta-analyses indicated activation divided by the total number of areas in which either one of two meta-analyses indicated a response. In this example, a reliability score of 42% is obtained. However, this may be an overestimate because several of the areas identified in these meta-analyses seem to be associated with processes that may not be specifically associated with the *n*-back procedure but, instead, with some other cognitive process. Of the five BAs reported to be common to both meta-analyses,

Table 3.6
A comparison of two meta-analyses of the brain activations associated with an *n*-back test

BA	Meta A	Meta B
6	X	X
7	X	
8	X	
9	X	
10	X	X
13		X
19		X
32	X	X
40	X	X
45	X	X
46	X	
47	X	
Total BA	12	
Common BA	5	
%	41	

Meta A shows data from Owen, McMillan, Laird, and Bullmore (2005), and Meta B shows data from Glahn et al. (2005).

one (BA 6) is a motor area; one (BA 40) is a "speech" area; and one is a part of the frontal cortex (BA 10) that is well known to respond to virtually any kind of cognitive activity. These three areas, therefore, may be generally associated with many other aspects of cognitive activity (e.g., responses) and may not be specific to the *n*-back test of working memory being examined here. On this basis, this comparison also seems to lead to the not unreasonable conclusion that the two meta-analyses do not reproduce the same pattern of activations; that is, they are not reliable indicators of working memory.

Another discrepancy between these two meta-analyses existed in the single non-Brodmann area mentioned. Owen et al. reported, ". . . there was a focus of activation in the medial cerebellum" (p. 50); Glahn et al. did not mention the cerebellum in their sample of healthy control subjects.

Whether the lack of consistent correspondence between the two meta-analyses is due to some subtlety of the method, difference in the tasks used in the original sample of experiments, arbitrary differences in the threshold for accepting activation, or some other cryptic source of bias cannot be determined now. Clearly, however, the two meta-analyses are not in agreement. The main point that they both seem to make is that there is nothing more than a substantial frontal commitment to working memory. Beyond that generality, there appears to be little agreement between these two meta-analyses.

Working Memory in General
Working memory is of such great interest to cognitive neuroscientists that it is unusual in providing many more opportunities for comparing meta-analyses than just the two I was able to find for the previous pairwise comparison. Indeed, I was able to find seven meta-analyses carried out on this cognitive process in addition to the two that used the n-back procedure. Of course, there were many different analysis techniques, responses, and, of course, a very different sample of selected studies used by each of these meta-analyses. Nevertheless, it is instructive to compare all of them to see if any consistencies might become apparent in this larger sample. Table 3.7 is a spreadsheet of the seven meta-analyses showing which of the 47 BAs were reported to be activated in each of these nine meta-analyses.[19]

A number of interesting observations can be gleaned from this spreadsheet. First, it is clear that there is a considerable amount of variability among the responding BAs reported in each of these meta-analyses. The extent of this variability is not surprising and can be partially explained by the variety of cognitive processes other than working memory intruding into the analysis. For example, if we just look at the BAs designated as activated by all or all but one of the meta-analyses, we see that only BAs 6, 10, 32, and 40 satisfy this rather loose criterion of commonality. However, as we have seen earlier in several instances, these areas are thought to be involved in brain cognitive activities that are not specific to working memory. For example, BA 6 is considered to be part of the motor system consisting of the premotor and supplementary motor cortices. BA 6, therefore, presumably would be involved in executing or planning the execution of motor responses, a necessary part of any measurement of cognition but not exclusive to working memory. Similarly, BAs 10 and 32 are frontal areas involved generally, as we see elsewhere, in virtually all cognitive processes and thus may not be specific to working memory. BA 40 is suspiciously close to what has traditionally been called Wernicke's

Table 3.7
A comparison of seven meta-analyses for various versions of the working memory task

BA*	Meta A	Meta B	Meta C	Meta D	Meta E	Meta F	Meta G
3							X
4				X			
6	X	X	X	X	X	X	X
7	X		X		X	X	X
8	X				X		
9	X		X		X		X
10	X	X	X		X	X	X
11	X						
13	X	X					X
17	X						
18	X						X
19	X	X	X	X			X
21	X			X			
22	X		X	X			
24	X					X	X
31	X						
32	X	X	X		X	X	X
37	X		X	X			
39	X						
40	X	X	X		X	X	X
44	X		X		X		X
45	X	X			X	X	X
46	X		X		X	X	
47		X				X	

*Plus scattered reports of thalamus, cerebellar, etc., activations.
The activation data in this table are taken from the following articles by column: A, Wager and Smith (2003); B, Glahn et al. (2005); C, Simmonds, Pekar, and Mostofsky (2008); D, Turkeltaub, Eden, Jones, and Zeffiro (2002); E, Owen, McMillan, Lair, and Bullmore (2005); F, Chein, Fissell, Jacobs, and Fiez (2002); G, Krain, Wilson, Arbuckle, Castellanos, and Milham (2006).

area and thus might have a more general language role than in terms of a specific working memory. Thus, depending on the design of the experiment, variability could be introduced.

We have to conclude from this examination of this extended group of meta-analyses that there is relatively little agreement among them that cannot be otherwise explained either by the activity of regions that have other sensory or motor functions or regions that are generally activated by almost any thought process. There is nothing indicated in either the paired or the eight-way comparison that suggests a common and specific indicator for any of the cognitive processes studied. Therefore, meta-analyses might have no probative value in linking brain-image responses and cognitive processes.

Additional Narrative Meta-meta-analyses

In the previous comparisons, I mainly exploited the advantage of comparing meta-analyses that reported their data in the form of the Brodmann system spatial locations, only adding non-Brodmann data when it was available. In addition, a number of other potentially comparable meta-analyses were occasionally encountered that provided their findings solely in the form of narrative descriptions of non-Brodmann locations. In the following pages, I look at a number of articles in which the activation locales were only described with narrative expressions such as "frontal" or "anterior cingulate" to define brain regions. Although this nomenclature is not as precise as the BA one, it is possible to get a qualitative estimate of how well two such narrative meta-analyses compare. As we now see, such comparisons also display a persistent and frustrating lack of reliability.

A pair of such narrative meta-analyses has been reported by Valera, Faraone, Murray, and Seidman (2007) and Ellison-Wright, Ellison-Wright, and Bullmore (2008), respectively. Their concern was not with functionally defined regions of activation but with size changes in the brain's anatomy that could be associated with attention deficit hyperactivity disorder (ADHD). Although the available sample of selected articles was relatively small in both meta-analyses and each used a different analytic procedure, it is instructive to compare the findings of these two structural meta-analyses because they both are trying to answer a very specific question: Did ADHD patients have different-sized brain parts than normal subjects?

Valera and her colleagues (2007) reported the following regions to exhibit reduced size in ADHD brains compared to normal controls:

The cerebellum

    The posterior inferior vermis

    The splenium of the cerebellar cortex

Total and right cerebral volume

Right caudate

Work from the Ellison-Wright laboratory (2008), however, reported size changes in ADHD patients in the following regions:

The right putamen

The globus pallidus

Obviously, there is little agreement between these two meta-analyses. Interestingly, their respective meta-analyses also showed that the statistical significance of several individual reports of increased size in a number of gray matter regions disappeared during their meta-analysis.

Although both of these studies can be considered to be very preliminary, it is clear that the two meta-analytic approaches produce very different initial answers to the question of brain component size differences between normal controls and ADHD patients.

To even further complicate the matter, I was also able to find another meta-analytic study of ADHD, albeit a functional rather than an anatomical one. This study (Dickstein, Bannon, Castellanos, & Milham, 2006) reported a very different pattern of activations than the regions implicated in the anatomical ADHD studies. Specifically, differences between control and ADHD subjects were reported in the following regions:

Frontal lobe

Cingulate gyrus

Parietal lobe

Basal ganglia/thalamus

Occipital lobe

Dickstein et al. paid special attention to the frontostriatal and frontoparietal circuits as potential "causes" of ADHD. Although it is not appropriate to compare the anatomical and functional characteristics directly, it is of interest to note that there is virtually no overlap among the regions implicated in the functional and anatomical meta-analyses, respectively. The only general inference one might draw is that widely distributed locations on the brain are implicated by all three studies.

In addition to the previously discussed work on emotional face recognition, the neural foundations of emotion itself have long been an object of theory. It was hoped early on that functional brain imaging studies might parse some of the complex interactions between the diverse brains regions involved in various kinds of emotion. Strategic approaches have varied over the years. Some investigators simply accept the current theoretical view that a particular region such as the amygdala or the frontal lobes is central to emotional behavior and direct their meta-analytic attention to its responses. Work by Baas, Aleman, and Kahn (2004) and by Sergerie, Chochol, and Armony (2008) are examples of this approach; both groups were mainly concerned with the laterality of amygdala function. Even within the constraints of this relatively simple question, diametrically opposite meta-analytic answers were obtained. For example, based on their meta-review, Baas, Aleman, and Kahn concluded that ". . . the left amygdala is more often activated than the right amygdala, suggesting different roles for the left and right amygdala in emotional processing" (p. 96).

Sergerie, Chochol, and Armony, however, came to what may be a very different meta-analytic answer to the question when they said ". . . our results do not support a stronger right amygdala involvement in emotional processing, nor a hemispheric lateralization based on valence or sex" (p. 823).

This disagreement should be considered in light of the fact that the Sergerie et al. (2008) meta-analysis summarized the results from 148 published experiments—an unusually large number for a modern meta-analysis. Similarly, the Baas et al. (2004) meta-analysis included data from 54 published reports.[20] Despite these relatively large databases, the two studies did not converge on the same answer; indeed, they provided diametrically opposed answers to what should be a very direct question.

Obviously, an enormous amount of effort has been expended in carrying out these meta-analyses. Unfortunately, this effort has resulted in inconsistent answers to the straightforward question of lateralization of the amygdala. Considering the variability that was exhibited in the individual studies and the fact that even these large meta-analyses did not resolve the issue, extreme care should be given to the credibility of any single experiment as well as the meta-analyses themselves.

Other investigators, however, eschewing an a priori commitment to the amygdala as a central part of the emotional system, have sought answers to the more general question: What parts of the brain are involved in emotional behavior? When the subject is framed in this way, the answers are much more complex if not equally inconsistent. For example, let us compare the answers to the general question provided by two meta-analyses published by Kober et al. (2008) and by Murphy, Nimmo-Smith, and Lawrence (2003), respectively. Each of these meta-analyses was founded on the basic premise of the meta-analysis approach, namely, that pooling the results of many individual reports would produce a more accurate estimate of the regions of the brain participating in a particular cognitive process. Nevertheless, despite all of the energetic pooling and statistical processing and our hopes and expectations of convergence onto a single, dependable answer to this question, these two reports differ substantially in their answer to the general question. Nor, for that matter, is either one forthcoming with results that are in accord with traditional brain models of emotional representation (e.g., as provided in the original version by Papez, 1937) or as modified by recent research (see Uttal, 2011, for a discussion of the evolution of the Papez circuit over the years).

Kober et al. (2008) list the following cortical and subcortical regions that their meta-analysis suggests is involved in emotional behavior. This summary is a result of their meta-analysis of 162 individual experiments.

Cortical Regions
- Dorsal medial prefrontal cortex
- Anterior cingulate cortex
- Orbitofrontal cortex
- Inferior frontal gyrus
- Insula
- Occipital cortex

Subcortical Regions
- Thalamus
- Ventral striatum
- Amygdala

- Periaqueductal gray
- Hypothalamus

An important outcome of the Kober et al. (2008) meta-analysis was the complex pattern of interactions among these regions. This was indicated by very widespread activity across the whole brain in emotional situations.

The most comparable meta-analysis in which the general question—What parts of the brain are activated in emotional behavior?—was addressed was that provided by Murphy et al. (2003). Also based on a large sample of individual experiments (106 published reports), they identified not only the main brain regions associated with emotion in general but also the specific emotions associated with each region. The result of their meta-analysis included:

Cortical Regions

- Insula/operculum (disgust)
- Lateral orbitofrontal (anger)
- Rostral supracallosal anterior cingulate (happiness and sadness)
- Dorsomedial prefrontal (happiness and sadness)

Subcortical Regions

- Amygdala (fear)
- Globus pallidus (disgust)

If we apply the same reliability score (the number of common areas divided by the total number of activated areas), we arrive at the value 42%. Four of the five common regions are located in the cortex; however, there are gross discrepancies between the subcortical regions associated with emotion by these two meta-analyses, the amygdala providing the only region cited in both studies.

Unfortunately, the narrative anatomical nomenclature used in making these comparisons is so poorly defined that it is not always clear what regions are being compared. For example, are the lateral orbitofrontal and the orbitofrontal regions of the brain really close enough to be considered to be a "common" location given the relatively large size of the frontal regions? Similarly, the large size of the deeply buried cingulate cortex makes it questionable how synonymous the rostral supracallosal anterior cingulate and anterior cingulate activation areas really are. Indeed, the breadth of the responses (ranging from the frontal to the occipital lobes) of the brain reported in these meta-analyses suggests a more widely distributed responsiveness, if not a brainwide one. If this is so, it may be that the emotional regions reported in these meta-analyses are just a sampling of the true extent of response during emotional cognitive activity, the highest peaks being separated from more generalized activity by a relatively high threshold of acceptance.

It seems clear that whatever disadvantages were present when we used the Brodmann system of relatively well-defined regions, they are compounded when we use this narrative nomenclature. Even with the most liberal interpretations of the constraints imposed by the narrative approach to localization, there appears to be little convergence on a common answer to the question of the locus of the emotional regions of the brain beyond a general frontal involvement. The lack of agreement is especially pronounced with regard to the subcortical regions.

Another area of definitional softness resides in the weakness of the psychological vocabularies. Murphy et al. (2003) parsed emotion into several categories based on the words that are used in common language. Kober et al. (2008), as well as Lindquist et al. (in press) on the contrary, dealt with emotion as a unitary entity. Again, it is appropriate for me to remind my readers that there are no robust a priori reasons why the psychological language with its arbitrary and ill-defined cognitive modules should map directly on neuroanatomical regions of the brain.

I make no claim that the comparisons between meta-analyses made here are any more definitive than the meta-analyses themselves. My comparison techniques are as frail as is the entire idea of pooling results from a number of experiments across the many factors that might influence their outcomes. The important point is that this first-order approximation, this first glance at the data, does not seem to support the basic meta-analysis postulate, namely, that by pooling data from a number of experiments, one is likely to converge on a better, more consistent answer to the localization problem. What we have shown here is that this assumption is not valid—multiple meta-analyses using a collection of inhomogeneous data do not converge on the same answer. Thus, at the present time, no meta-analytical finding should be considered to be a valid statement of which brain regions should be associated with which cognitive processes.

The next section presents an overall view of the implications of this pattern of inconsistent results.

Box 3.3
Empirical Conclusion Number 3

> Alternative meta-analytic summaries of many experiments purporting to find relations between a cognitive state and a pattern of brain responses do not agree with each other. This raises serious questions about the validity of the idea of pooling results from a group of experiments in order to provide a more accurate estimate of the activation pattern than is possible from a single experiment. The subtle clustering that can be teased out of the data by a meta-analysis may be due to random factors or the demands of the analysis itself instead of real effects due to the cognitive process under investigation.

3.2.4 What Does This Pattern of Results Mean?

To summarize the results of this chapter, inconsistencies abound at all levels of data pooling when one uses brain imaging techniques to search for macroscopic regional correlates of cognitive processes. Individual subjects exhibit a high degree of day-to-day variability. Intersubject comparisons between subjects produce an even greater degree of variability. When responses from a group of subjects are pooled into experiments, seemingly comparable experiments produce inconsistent indicators of which portions of the brain are associated with particular cognitive processes. Finally, as we have seen, when we compare meta-analyses in which the findings from multiple experiments are selected and their outcomes pooled, we also find substantial disagreement.

The overall pattern of inconsistency and unreliability that is evident in the literature to be reviewed here suggests that the intrinsic variability observed at the subject and experimental levels propagates upward into the meta-analysis level and is not relieved by subsequent pooling of additional data or averaging. It does not encourage us to believe that the individual meta-analyses will provide a better answer to the localization of cognitive processes question than does any individual study. Indeed, it now seems plausible that carrying out a meta-analysis actually increases the variability of the empirical findings. The Simpson paradox is a strong argument that meta-analyses, however promising they may initially seem to be, have not yet proven that they generally work.

To appreciate what this means we have to reconsider the basic philosophy behind the meta-analysis approach—the idea that pooling of data from many experiments can help us toward a more precise and specific answer to the question of localization of cognitive function. This was supposed to be effected by increasing the power of statistical analysis, presumably by incorporating more and more data into a very large virtual sample. The proposed process was supposed to be analogous to simple averaging of signals to isolate a repetitive signal from the variable background noise. In the case of brain imaging technology, the idealized signal being sought is supposed to be a precise answer to the subtle localization of a cognitive process, and the noise is the result of the substantial variation in selecting subjects, stimuli, experimental design, responses, and all of the other factors that can influence a brain image (see especially table 2.1). By pooling data from a number of experiments, the presumption is that this noise could be minimized and the resulting meta-analysis would provide a robust and statistically significant specification of the brain areas associated with whatever cognitive process is under study.

We see in this book many indications that suggest that these assumptions are yet to be confirmed. I have already pointed out the tenuousness of the idea that two or more spatial regions can be combined in some way to produce a meaningful average region. (This was referred to as the Venn fallacy in chapter 1.) It is also yet to be established whether combining wildly inhomogeneous brain imaging data is a valid statistical enterprise. Not all of the findings that make up the meta-analysis are associated, for example, with a common effect dimension. Nor, for that matter, are the criteria for accepting or rejecting activations consistent from one experiment to another. Finally, there are a host of technical and selection issues that also exacerbate the lack of homogeneity of the pooled data, thus reducing the reliability of the method.

The point is that by simply dumping all of these experimental findings into a common statistical pot, without adequate consideration of the diversity of common inputs, attributes, and effects, no amount of numerical manipulation in even the best-designed meta-analysis is likely to produce consistent results. However, no amount of philosophical argument, no matter how logical, is going to put this issue to rest. The proof is in the pudding, as they say (the pudding in this case being the reliability and replicability of the empirical findings). That is, as it is with any other scientific endeavor, the ultimate test of the robustness of the meta-analytic approach is to be found in its consistency from situation to situation.

Therefore, the empirical question is: When we compare two meta-analyses, do they agree? The general answer to this question, based on the comparisons presented in this chapter, seems to be *no*—they do not agree with the degree of confidence that would lead us to infer that they were reliable indicators of mind-brain relationships!

This brings us face to face with the necessity to understand why there are inconsistent answers emerging from comparable meta-analyses. There are several possible explanations for this unreliability. The first possibility is that despite efforts to select relevant studies, the selected experiments involved in each of the meta-analyses differ so much that it was not really appropriate to pool their findings. This is an extension of the "apples and oranges" inhomogeneity argument against meta-analyses that has pervaded the field since the technique was first proposed. This explanation of the observed inconsistency is based on the premise that the properties of the pool of experiments combined in a meta-analysis actually varied too much among themselves to be combined by such a technique. Although each of the meta-analytic research teams may have used seemingly reasonable selection criteria, they actually may have combined experiments that were in fact so different that the fallible selection process doomed the enterprise from the beginning.

Why should this be the case? A major reason that psychology may have been especially susceptible to such selection errors, as I have noted several times, is the vague way in which we define psychological processes. Two experiments that purport to study working memory may actually be evoking quite different cognitive processes because of some subtle (or not so subtle) task difference. In this context, once again, we are confronted with the limitations of the language used by psychologists; what appear to be synonyms may not, under careful scrutiny, turn out actually to denote the same kind of cognitive process. Furthermore, identical cognitive process may be classified with different names and thus incorrectly be excluded from a meta-analysis. According to this explanation, any differences between or among the meta-analyses may be accounted for in terms of the weakness of the psychological vocabulary.

A second possibility explaining the failure of meta-analyses to agree is that the analysis method used to produce each meta-analysis differed in some subtle, but critical, way. However, this is not likely because most of the meta-analytic studies compared here used the Turkeltaub, Eden, Jones, and Zeffiro (2002) ALE method.

Variation in the arbitrary thresholds for acceptance of an activation chosen in the original experiments may also have subtly contributed to a lack of homogeneity in the pool of experiments, thus obscuring any commonalities—if, in empirical fact, such commonalities actually existed. This problem is exacerbated by the all-or-none scoring of regional activity (i.e., simply indicating the presence or absence of a response) in all brain imaging studies.

Another possibility is that the signal-to-noise ratios characterizing the results of the pooled experiments were simply too poor to permit the extraction of any commonalities from the original brain imaging data. Such a source of bias would be enhanced by the fact that noise levels could have been quite different from one experiment to another. The failure in this case would be in the meta-analytic methodology. It is possible that the methodology we are using is simply not powerful enough for the task at hand—associating the macroscopic spatial responses of the brain to cognitive processes. The whole brain's coding mechanisms may be too subtle and complex for the relatively primitive techniques used to combine the outcomes of a group of experiments.

A related possibility explaining the absence of agreement among the meta-analyses is that the salient and important data were lost somewhere in the complexities of the necessary analyses, pooling, and data combination. It would not be easy to determine where this loss of critical signal information occurred. However, one possibility is that at some level of the sequential series of accumulations and pooling of data from subjects and then experiments, the signal-to-noise relations were degraded rather than enhanced. It could have been, for example, because the pooling of data heaped asynchronous noise on asynchronous noise in a way that acted to enhance the noise while diminishing the signal contrary to basic assumptions of the meta-analysis approach.

According to this explanation, as the pooling level increased and the averaging progressed, the signal-to-noise situation progressively worsened rather than improved as expected. An increase in apparent randomness with progressive pooling was presciently noted by Cox and Smith (1953, 1954) when they described how regular interval patterns between multiple spike action responses from a group of neurons quickly approached a kind of quasi-randomness when they were pooled. This phenomenon occurred even when the most orderly and periodic signals were pooled. In a far simpler case than the brain images being considered here, they were able to show that progressive pooling led to an analogous kind of information loss that may possibly account for this progressive divergence of responses during a meta-analysis.

A far more serious explanation for the future of brain imaging is that the processes and mechanisms under study *are* essentially random; that is, a common, deterministic signal of a cognitively significant pattern of brain responses actually does not exist! That is, despite 20 years of intense research, there may have been no localized brain signal to extract from whatever amount of noise may have been present. In other words, the lack of replicability of brain image findings at all levels of pooling may indicate that these macroscopic brain image signals may be, for all practical and theoretical purposes, quasi-random, if indeed not fully random! The poor signal-to-noise ratios and the erratic nature of the data may have permitted cognitive neuroscientists to "see" order where there is, in actual fact, none.

The implications of such a conclusion are profound for cognitive neuroscience. It would suggest that the entire enterprise of searching for macroscopic brain correlates of a cognitive process might be a search for a nonexistent chimera. From a certain point of view, this negative conclusion would still be a major contribution to cognitive neuroscience: it would shift attention and support away from what may be an erroneous line of research at the wrong level of analysis. Beyond that, if the kind of inconsistency I have pointed out here turns out to be the norm, we may have witnessed a serious misdirection of effort and resources in the field of cognitive neuroscience.

Any author of such a drastic conclusion as the one presented here—brain images may be artifacts and may not be significantly correlated with cognitive processes—must leave open the challenge to anyone who disagrees—do your own comparisons and meta-meta-analyses to see if you can demonstrate a more robust form of consistency and reliability.

No one can predict the scientific future. It is always possible that future improvements in methodology might be able to find regularities where none can be detected now. However, at the present stage of development of cognitive neuroscience and the tools it uses, the kind of reliability and replicability we need to draw robust conclusions about the mind-brain relationship is absent. At the present moment, however, the conclusion that one must come to after surveying the empirical literature and examining it for reliability and consistency can be summed up in the general statement presented in box 3.4.

Box 3.4
Empirical Conclusion Number 4

> Given that reliability is low at all levels of analysis (intrasubject, intersubject, interexperiment, and inter-meta-analysis), it is increasingly likely that relations between particular macroscopic brain images and specific cognitive processes have not yet been established and may not exist. The many reports that purport to having established such relations are possibly reporting random or quasi-random fluctuations in extremely complex systems inadequately controlled by current experimental paradigms.

4 Macroscopic Theories of the Mind-Brain

4.1 Introduction

The modern development of complex instrumentation and mathematical methods for studying the relationship between cognitive process and the brain has skyrocketed cognitive neuroscience to a position of prominence in both funding and the popular media. The near-universal interest in what neural processes and mechanisms account for the richness of our mental lives has resulted in a barrage of theories and interpretations that sometimes seems to outrace the supporting empirical evidence. This is particularly evident in the context of chapter 3 in which we demonstrated the inconsistency and unreliability of many of the findings from what seems to be an exponentially expanding body of results.[1] Not only were the findings from individual subjects and from similar experiments different, but even when high-powered meta-analyses were brought into play, inconsistency seemed to be characteristic of even the most basic results.

Why this is the case is not too difficult to understand. The basic problem is the complexity of the system under study. Because of this complexity, there are a large number of potential biases and errors of experimental design and analysis that contribute to making brain responses so variable. Notwithstanding the undoubted utility of brain imaging devices as tools for studying anatomy and physiology in many fields of biological science and medicine, the inconsistency of the data they produce in cognition-related experiments should at least raise the question of their utility in that context. It is possible, although I expect that few researchers in this field would agree with this assertion, that with only a few exceptions, modern brain image comparisons with cognition have not yet fulfilled the hopes of cognitive neuroscientists.

Before considering various theoretical approaches to brain-cognition systems, it is important to reiterate a particular point about the general relationship between these macroscopic brain regions and cognitive processes. Prevailing opinion nowadays is that the psychoneural equivalents (i.e., the neural processes that actually account for, instantiate, represent, encode, or otherwise provide the neural basis for cognition in all of its manifestations) are to be found in the complex network of neurons that make up the brain. Unfortunately, for several reasons we are prevented from examining mind-brain relations at this microscopic level. The main reason, of course, is the sheer numerousness of the neuronal networks—numbers that are so large as to prevent us from examining the actions and interactions of literally billions of simultaneously and idiosyncratically behaving neurons by any conceivable process. Another reason is that the brain imaging instrumentation available to us, with all of its wonderful attributes, is essentially a very specialized tool for examining the macroscopic level of brain regions, centers, and nuclei—the spatial and cumulative properties of the system—and not what is more likely to be the truly germane microscopic level.

Therefore, both sheer numerousness and available technology drive us to direct our research attention to what may be an irrelevant level of analysis—the macroscopic properties of the brain. As a result, although there may eventually be some kinds of simple correlations or biomarkers discovered between macroscopic brain activity and cognition, this is not the level at which the most important questions should be asked. The microcosm may be where we should look; because we cannot, we attack the problem in terms of the current context of what we can do. Unfortunately, by doing so, we may be directing ourselves along the wrong path.

This perspective is based on the idea that there are really two levels of theory, one microscopic, and one macroscopic, that do not speak directly to each other. The degree of microscopic order or disorder that is considered here (that of the 10 billion or so neurons of the brain) is unknown—it is conceivable that the mind-brain system is so complex that researchers will never be able to disentangle the cognitive and neurobiological links, correlates, and interactions that underlay mental activity. Neither the powerful human pattern perception capabilities nor those of the most powerful computers of which we may conceive are capable of working their way through the huge numbers involved in the irregular neuronal networks

Macroscopic Theories of the Mind-Brain

that make up the essential properties of the brain. This irregularity is not randomness; the mind-brain system cannot be considered to be random in any formal sense. The simple fact that we function, learn, perceive, and behave in such an adaptive manner belies any attempt to designate the mind-brain as a truly random system. Instead, what we are dealing with is a kind of pseudo- or quasi-form of approximate randomness that arises from the complexity of the neuronal network that instantiates mind.[2]

Despite this quasi-randomness cum complexity, cognitive neuroscientists make persistent efforts to explain, interpret, describe, model, or theorize about what their data are telling them. This is what the best of science is supposed to do. Raw data mean very little in themselves; extrapolating from the data to develop an explanatory model is the essence of the crown jewels of science such as Newton's and Einstein's physics, Mendel's genetics, Mendeleev's chemistry, as well as Linnaeus's and Darwin's biology. Data do not speak for themselves; they are intrinsically meaningless and must be interpreted in a way that leads to as much synoptic theory as possible. Indeed, any argument that empirical data obtained from carefully structured and rigorously controlled experiments constitute the prime goal of science is not sustainable. The ultimate goal of any science should be the summarizing and consolidation of narrowly construed data into general and comprehensive theories. It is the synthetic interpretation of the results of analytic experiments into synoptic theories that is the ultimate goal of a science.

The holy grail of theory in cognitive neuroscience is the answer to the question of how the brain makes the mind. I might as well confess that which everyone knows. There currently is no theory or explanation at at any level of analysis of how the brain makes the mind beyond the most constrained and vague speculations! Furthermore, because of the aforementioned complexity, there may never be. Our theories of the mind-brain relationship are limited to relatively minor aspects of the grand question itself.[3] We know a little about sensory and motor coding, a lot about how individual neurons function, and have a lot of questionable knowledge about how the brain operates as a whole when we are thinking. The best we can currently assert is that whatever the mind is, it is a function or process of the neuronal networks that make up the brain.[4] Without at least a tacit acceptance of the material physicalism implicit in this assumption, this science could not exist. Who could possibly carry out any experiment in cognitive neuroscience constantly shadowed by the possibility that unknown extraphysical (i.e., dualistic) forces were at work to modulate or confound the controls built into our experimental protocols.

However convinced one may be concerning the role neuronal interconnections play in creating mental activity, it is important to also appreciate that there is still not a shred of scientific evidence that mind emerges from activities at this microscopic level of analysis. It is only a "best bet"—a speculative hypothesis—that we are probably forever prohibited from probing because of the enormous complexity and numerousness of the brain interconnections. The fact that logic, reason, and objective scientific evidence have not been able to come up with any compelling alternative remains the strongest argument for the neuronal network hypothesis of the origins of the mind. However much truth there may be in this form of argument—proof by exclusion—this is not the way that scientific theory building is supposed to operate—such theories are always vulnerable to the next piece of evidence. Unfortunately, we are left with macroscopic approaches (e.g., the brain imaging and EEG techniques) that pool and obscure the details of the truly salient neuronal networks.

Another important issue in our discussion of brain organization theories is the spatial nature of the data obtained from brain imaging. That is, with very few exceptions, the application of these powerful methods can provide answers only to where activation responses occur. The science is relatively limited to questions such as "What regions of the brain are activated during a particular cognitive process?" or its inverse, "What are the cognitive functions of a particular brain region?" In relatively few cases is there any attempt to quantify the magnitude of an effect (the probability that a response pattern is different from a control condition is not a measure of the amplitude of an effect). Similarly, the temporal dynamics of the brain's responses plays a minor role in most contemporary research and theory.[5] Therefore, virtually all of the emerging interpretations of what has been found are theories of what part or parts of the brain are supposedly involved in the cognitive task being studied.[6] Although the brain imaging approach is intended to tell us specifically "where" something may be happening, the most recent findings seem to be telling us "almost everywhere." In such a context, theory building is very difficult, indeed.

Finally in this introduction, I should note that the field of brain imaging from its start has been heavily committed to the mathematical and statistical aspects of its methodology. Although this is a desirable direction for some of the science's efforts to have taken, there is some suggestion that this is also a diversion in which the "tools" of the science are polished and polished again in the hopes that some new manipulation will be able to better extract ever more subtle results than were hitherto forthcoming. Some might refer to this retreat to methodology as displacement activity; when one cannot find order in the data, one seeks to sharpen one's available tools in the hope that an enhanced method will uncover some hidden significance. Unfortunately, as we see repeatedly evidenced in chapter 3, the most elaborate and up-to-date methods (e.g., the ALE system of Turkeltaub et al.) still do not provide reliable and consistent answers to the many questions that cognitive neuroscience currently asks.

Of course, it is always possible that the hope that future developments in methodology will eventually be fulfilled by techniques with greater sensitivity and stability than those available nowadays. However, it is also possible that no future developments will ever be able to parse the kind of complexity observed either at the macroscopic or microscopic level of analysis of the brain. The lack of reliability of the responses from the single subject to the most comprehensive meta-analysis may signal the kind of quasi-random character of brain imaging findings that will continue to bedevil modern theory building. If failures of reliability continue to characterize future research, it is unlikely that any amount of "tool sharpening" will ever be able to make this pig's ear of inconsistent data into the silk purse of valid explanation of how our cognitive processes might emerge from macroscopic chunks of brain tissue.[7]

This, then, sets the theme for this chapter. In it, I examine the modest kinds of theories that cognitive neuroscientists currently infer from their data. First, I present a preliminary taxonomy of theoretical types; then I survey the literature to determine how these various types have been used by modern cognitive neuroscientists to support their favorite mind-brain theory.[8] I then turn to a number of special topics concerning the current state of theory in cognitive neuroscience.

4.2 Types of Macroscopic Theories of Mind-Brain Relationships

Given that almost all of the data from brain imaging experiments are spatial—that is, answer to the "where" question—it is not surprising that most of the theories and interpretations of those data are also spatial in nature. However, that does not mean that all such theories are identical. To the contrary, over the years there have been a variety of different interpretations of the data forthcoming from brain imaging experiments. The differences arise from a number of sources, most prominently the changing nature of the data themselves but also including changes in the goals of the investigators and, of course, the availability of new measuring instruments. The following categories suggest the divergent approaches to conceptualizing the relation between the mind and these kinds of neurophysiological data:

- Biomarkers
- Localized nodes
- Distributed systems with function-specific nodes
- Distributed systems without function-specific nodes

4.2.1 Biomarkers

Most of the investigators active in the biomarker field eschew any particular theoretical concerns with the classic mind-brain problem. Their concerns are primarily therapeutic or diagnostic; their goal is simply to find some correlated activity that they can use to measure or predict the presence or course of a disease state. Any signal that correlates with the dysfunctional cognitive state is acceptable to fill this role; it need have no explanatory or even descriptive relevance. The use of brain images as biomarkers, as with any other neurophysiological measure, is simply to have available a correlate of the dysfunctional condition to help in its detection, diagnosis, or prognosis. That there may be no known causal link between the neurophysiological signal and the cognitive process does not concern seekers of biomarkers. To the extent that such a signal can be useful, there is no reason that such an atheoretical approach should be looked on with disdain by investigators with more arcane goals.

The concept of the biomarker as a clinical tool for diagnosis and prognosis prediction is implicit in much of the history of medicine, especially in the field of psychiatry. However, it is not a traditional way of thinking in a cognitive neuroscience that wishes to explain and theorize about causal relationships. The idea that there may be utility in the use of a physiological response as a biomarker, as opposed to the definition of a causal factor, nevertheless, has grown over the last few years. Singh and Rose (2009), for example, reported the increasing use of the concept in the psychiatric literature in a chart originally attributed to J. A. Rached of the London School of Economics and Political Science.

Why the biomarker idea has grown so rapidly is not difficult to discern. The most obvious reason is that many more potential biomarkers are currently available to be compared to cognitive processes than in previous years. The growth of biomarker-related publications overlaps quite well with the increase in brain imaging techniques, if slightly delayed.

Although there are many scientific (a biomarker may give a false sense of understanding of the true roots of a disease) and ethical problems (should biomarkers be used to predict future diseases?) associated with their use, it can without doubt satisfy some useful needs in some cases in which the actual physiological causes are either unknown or unknowable. (The most familiar biomarker illustrating this indirect role is, of course, the human body temperature.) In fact from a certain perspective, proponents of the use of biomarkers without the theoretical or causal accoutrements may be closer to the current state of the art than are the more arcane cognitive neuroscientists. Much of current theoretical and causal thinking may actually be unsupportable; despite the hopes of those who seek causal relationships between brain images and cognitive processes, there may be in empirical and inferential fact nothing yet there beyond the kind of correlative associations highlighted by the biomarker community.

The range of possible applications of the biomarker concept is enthusiastically summarized by Singh and Rose (2009):

To diagnose a condition; to predict the natural outcome for an individual with this condition; to predict whether the individual will benefit from a particular treatment and how aggressively to treat the individual; and to assess an individual's response to this treatment. In a psychiatric context, biomarkers could be used to detect and assess, or predict the development, not only psychiatric disorders, but also personality or behavioural [sic] traits, and emotional or cognitive capacity. (p. 204)

Although this may be stretching the point, the key idea is that biomarker indicators (correlates) may serve useful albeit atheoretical functions that make no attempt to determine the actual causal links of those biomarkers with cognition or behavior. Indeed, this limited role of the biomarker is rarely discussed. Such issues as the possibility that the behavior and the biomarker may be only indirectly related through the medium of a third cryptic factor driving both are typically overlooked.

Again, it must be reemphasized that even the most modest use of brain images as biomarkers for cognitive dysfunction is still without empirical validation. Despite the vigorous search for lie detectors or aggression-correlated biomarkers, such indicators do not yet meet modern standards for scientific acceptance. This, of course, does not mean that the search has ceased. The most extensive historical example of the unsuccessful search for clinically useful biomarkers is probably in the study of schizophrenia. Nowadays, such investigators as Javitt, Spencer, Thaker, Winterer, and Hajos (2008) have argued that the use of a variety of modern biomarkers (including EEG, P-300 ERP, and fMRI) can be useful in studying the effects of drugs in the treatment of this incapacitating disease. This is so despite their assertion that ". . . underlying genetic and neuronal abnormalities are largely unknown" (p. 68). Even more fundamental, however, is the fact that reliable biomarkers for schizophrenia do not yet exist. Javitt and his colleagues are expressing a hope, not a fulfilled quest. Although the importance of finding a valid biomarker for schizophrenia has stimulated an extraordinary amount of research for decades, even the most promising leads tend to fizzle out when research is replicated. The residual problem is what constitutes a "sufficiently high correlation" for diagnosis with all of its ethical and medical implications.

A similar perspective characterizes the search for biomarkers of another pair of serious cognitive dysfunctions—autism and ADHD. Both of these dysfunctions are clinically defined by abnormal behavior, and although there has not yet been any agreement on a neurophysiological or neuroanatomical cause of these dysfunctions, there has also been a concerted search to find some biomarker that will at least support diagnosis and prediction. Unfortunately, none has yet been found that could be considered to be an objective indicator of the cognitive dysfunction.[9]

Courchesne et al. (2007) recently reviewed the field, noting the variety of candidate neuroanatomical biomarkers that have been shown to correlate to at least some degree with autism. These included the size of certain parts of the brain, the thickness of the cerebral cortex, neuronal growth patterns, and head circumference. Unfortunately, some these markers could only be measured post mortem and, thus, were not very useful as predictors. Nor, for that matter, was there any explanation about how these anatomical irregularities might be associated with the behavioral symptoms of autism.

Once again, the modern availability of fMRI systems has opened the door for new approaches to identifying physiological as opposed to anatomical biomarkers. In another review, Levy, Mandrell, and Schultz (2009) list some of these proposed biomarkers of autism:

> Functional MRI has shown difference in patterns of activations and timing of synchronization across cortical networks, with lowered functional connectivity relating to language, working memory, social cognition or perception, and problem solving. The most reliably replicated functional MRI abnormal finding is hyperactivation of the fusiform face area, associated deficits in perception of people compared to objects. Results of other functional MRI studies done during imitation tasks have suggested impaired mirror neuron function in the inferior frontal gyrus (pars opercularis). (p. 1630)

However, they concluded that "Attempts to identify unified theories explaining core and comorbid deficits have been unsuccessful, which is not surprising in view of the heterogeneous expression of autism spectrum disorders" (p. 1630). In other words, none of the proposed biomarkers yet works!

Other researchers seeking biomarkers for such behavioral disorders as autism and ADHD have turned to biochemistry in their search for biomarkers. Bradstreet, Smith, Baral, and Rossignol (2010), for example, have reported the following medical conditions associated with either autism or ADHD by one or another study: ". . . oxidative stress; decreased methylation capacity; limited production of glutathione; mitochondrial dysfunction; intestinal dysbiosis; increased toxic metal burdens; immune dysregulation, characterized by a unique activation of neuroglial cells; and ongoing brain hypoperfusion" (p. 15).

Although the relation between these chemical conditions and the behavioral disorders also remains unknown, Bradstreet et al. suggest that these biomarkers should be treated "even if no clear immediate behavioral improvements are observed" (p. 15). This is clearly inconsistent with the theory-based approach of many cognitive neuroscientists. It suggests a shotgun approach that might be extremely wasteful of resources and potentially harmful to the patients, there being not even a hypothetical causal link between the biomarker and the behavior. The desirability of medically treating an abnormal biomarker in the absence of any knowledge of its relationship to the behavior is a major ethical and medical issue deserving considerable discussion.[10]

The situation is much the same for potential brain image biomarkers of ADHD. Paloyelis, Mehta, Kuntsi, and Anderson (2007) have extensively reviewed the literature and found that fMRI biomarkers based on differences between ADHD patients and a normal control group were reported in virtually every region of the cerebral cortex by one or another of 32 published articles. The discrepant and inconsistent nature of these data suggests that, although there may be some cryptic biomarker of ADHD hidden in some brain images, any diagnostic application of it must lay far in the future. Paloyelis and his colleagues suggest that the future is "five years"; some of us think that this may be a gross underestimation.

To generalize from these results, it seems that the search for adequately correlated biomarkers—a pretheoretical mode of research in a field that is characterized almost entirely by behavioral dysfunction—is currently still unfulfilled. A few neurophysiological responses and neuroanatomical structures in the brain have been shown to correlate with these maladies to at least some degree; however, the results are inconsistent and, to a large degree, questionable in terms of any direct causal linkage. The absence of any of these macroscopic indicators suggests, once again, that either this is the wrong level at which to seek a dependable biomarker of cognition or there is no biomarker to be found in brain images or brain chemistry for a very basic reason. The true neural equivalents of these behavioral disorders are subtle changes in the state of the neuronal network, not in the chunks of the brain or in the chemistry of transmitter substances that may control these changes. In fact, one might go even further and speculate that some of these behavioral disorders, no matter how debilitating, may represent exacerbations of what would otherwise be considered to be normal cognitive functions. If that is the case, then any hope that a macroscopic biomarker would be found for any of them appears to be increasingly remote.

4.2.2 Localized Nodes

Although it has so far proven to be difficult even to find a reliable biomarker of such subtle cognitive dysfunctions as autism, cognitive neuroscience researchers continued to seek tighter relationships between brain responses and mental activity. One might well ask, if we cannot find reliable biomarkers, what chance is there of unearthing signals that actually reflect causal relationships? Nevertheless, the search goes on and has been particularized in the last few decades in the debate between localized and distributed theories. The localized theory proposes that the neurophysiological responses to a cognitive task are constrained to narrowly circumscribed cerebral regions. This theory suggests that each part of the brain has some specialized and possibly independent function-specific role. On the other hand, the distributed theory proposes that the brain responses associated with particular cognitive processes are spread over broad swaths of the brain. Considering the current status of the evidence, it seems clear that the answer to this question must now be phrased in the terms of broad distribution. Only a few current investigators report unique, localized representations, and this is probably due to statistical manipulations (such as the use of too high a threshold) rather than to the realities of the response.

The traditional theoretical approach to the relationship of macroscopic brain regional activations and cognitive processes dates back to the phrenology of Gall and Spurzheim (1808). No one nowadays gives credence to their "bumps on the skull" theory, but the basic underlying assumption of what was at their time a very popular enterprise—that there are narrowly circumscribed regions on the brain (as reflected in the bumps on the skull) specialized to carry out specific cognitive processes—still motivates, albeit implicitly, many cognitive neuroscientists. The idea of distinct localized regions, each with a distinct cognitive function, persisted until the last decade's eruption of brain imaging studies that showed that brain activity was much more broadly distributed than would support any kind of an extreme localization theory.

The basic "phrenological" idea of localized brain regions representing specialized cognitive processes characterized the entire cognitive neuroscience enterprise for years prior to the development of PET and MRI systems. This particular history cannot be overlooked in any general historical view of the science. Its influence can be seen in the older techniques of experimental brain surgery as well as in the emphasis put on trauma cases. Experimental brain surgery and trauma cases both played the same game—determine what cognitive or behavioral dysfunctions occurred when a part of the brain was removed by intent or by accident. The prevailing theory was that when an association was found between a damaged region and a cognitive activity, that region was ipso facto the locus of the missing cognitive function.

Simplistic high-level cognitive neophrenology of this kind was abated by what clearly were observations of localized brain function in the sensory and motor domains. The discovery of the occipital visual area (Munk, 1881), the mapping of the somatosensory homunculus (Woolsey, 1952) and the auditory areas (Tunturi, 1952) as well as what seemed to be dedicated speech areas (Broca, 1861; Wernicke, 1874) were uncritically extrapolated to models assuming similar localized brain mechanisms for higher-level cognitive processes such as thinking, rage, or affection. However, as I have repeatedly pointed out, these sensory and motor input-output mechanisms differed in major ways from cognitive processes. Typically, they were elicited by well-defined physical stimuli, required simple discriminative judgments, and were probably better considered as transmission rather than representational systems. Therefore, we should probably consider them to be poor models of the much more complicated higher-level cognitive processes.

It was with the rush of new data from the brain imaging devices that these simplistic models of function-specific and localized brain representation of cognitive modules began to fall apart. The typical pattern of PET and fMRI responses to cognitive tasks turned out not to be a single or a few demarcated places on the brain but rather a multiplicity of poorly defined regions. Furthermore, no region was function specific; all had multiple functions. The basic empirical fact emerging from this kind of research was not a version of neophrenological localization but, rather, of a wide distribution of multiple activations.

Lindquist, Wager, Kober, Bliss-Moreau, and Barrett (2012), among others, also came to the conclusion that there is no basis for any kind of function-specific localization. They summarize their results concerning emotion with the following statement:

Overall, we found little evidence that discrete emotion categories can be consistently and specifically localized to distinct brain regions. Instead, we found evidence that is consistent with a psychological constructionist approach to the mind: a set of interacting brain regions commonly involved in basic psychological operations of both an emotional and non-emotional nature are active during emotion experience and perception across a range of discrete emotional categories. (p. 1 in preprint)

What this pattern of activations should have implied is that regardless of the cognitive process being considered, many if not most brain regions

—not a single one or a few—are activated. It now seems beyond reasonable contention that the brain's response during a high-level cognitive process is made up of a distributed system of many different multifunctional locales or regions or of large regions of the brain. The extent and the nature of this distribution are not yet definitively known for any cognitive process, but there is little doubt that the evidence for distribution heavily outweighs that for localization. This does not beg the question of what these responses mean in terms of the representation of cognitive process; it only adds further support to the argument that the old hypothesis of narrowly localized function-specific nodes can no longer be considered to have any validity. The amount of additional evidence to support this conclusion is now overwhelming. Distribution, not localization, must be the foundation of any future theory of mind-brain relationships.[11]

At this point, we consider two alternative theoretical directions that theory can take from this seemingly irrefutable empirical result. One, the system of nodes approach, preserves something of the neophrenological tradition by maintaining that although the brain response to any stimulus or task is broadly distributed, it can still be characterized as a system of discrete and localized nodes, perhaps corresponding to anatomical structures, each of which has a specialized function. The other, the "soft" distribution approach, considers the brain to be more of a boundary-free, undivided system without any function-specific localized regions other than in sensory or motor processing regions. As we see, the evidence to distinguish between these two kinds of distribution is not as compelling as that distinguishing between function-specific localized nodes and a distributed system without them. Indeed, the difference between a system of discrete nodes and a softly distributed system may also depend on an arbitrary judgment. The data are equivocal on this issue, and the difference may be made not so much based on the empirical data but, rather, on the theoretical orientation of the cognitive neuroscientist. The next two sections seek to clarify the distinctions between these two alternative views of brain organization.

4.2.3 Distribution with Function-Specific Nodes

The idea that different parts of the brain may perform different functions but must cooperate to encode or represent cognitive processes is not a new one. One premier example of this approach can be found in the work of Papez (1937). Papez was an anatomist who had been influenced by previous work in which a resolution of the James-Lange and Cannon-Bard controversy concerning the nature of emotions was sought.[12] Much of the earlier work (prior to Papez's contribution) on possible neural mechanisms of emotions had been directed at the hypothalamus—an example of the tendency to localize this kind of cognitive function within a single portion of the brain's anatomy. Papez, however, as an anatomist, understood that the hypothalamus was heavily interconnected with other portions of the brain. Based on his understanding of the anatomical interconnections, he concluded that the hypothalamus did not operate alone but rather that "Taken as a whole, [an] ensemble of structures is proposed as representing theoretically the anatomic basis of the emotions" (p. 725). In Papez's original system the "ensemble of structures" included the following brain structures:

- The hypothalamus
- The cingulate gyrus
- The anterior thalamic nucleus
- The mammillary bodies
- The hippocampus
- The subiculum
- The parahippocampal gyrus
- The entorhinal cortex

Subsequent research has made it clear that the frontal cortex and other regions of the brain also play major roles in this system. The essential point of Papez's theory was that the salient brain components of emotion were represented by a broadly distributed system of nodes. Indeed, as we have learned more and more about the brain responses, it was hard to find regions that are not involved in emotion.

Similar brain systems have been proposed for other cognitive activities. Thompson (2005), for example, proposed a conceptually similar system for declarative learning and memory that incorporates the following brain structures:

- Cerebral cortex
- Hippocampus

Macroscopic Theories of the Mind-Brain

- Cerebellum
- Striatum
- Amygdala

Johnson (1995) carries out the same task for visual attention. He lists[13] the following brain regions as involved in this cognitive process:

- Visual areas V1, V2, V3, V4
- Frontal eye fields
- Dorso lateral prefrontal cortex
- Basal ganglia
- Substantia nigra
- Lateral geniculate
- Medial temporal cortex
- Inferotemporal cortex
- Superior colliculus
- Brainstem

The key idea in this type of theory is that, although a part of a distributed system, the constituent nodes are localized and function specific to a particular cognitive process. In the words of Posner, Petersen, Fox, and Raichle (1988):

The hypothesis is that elementary operations forming the basis of cognitive analyses of human tasks are strictly localized. Many such local operations are involved in any cognitive task. A set of distributed brain areas must be orchestrated in the performance of even simple cognitive tasks. The task itself is not performed by any single area of the brain, but the operations that underlie the performance are strictly localized. (p. 1627)

Posner and Rothbart (2007) have renewed this assertion: "Results of neuroimaging research also provide an answer to the old question of whether thought processes are localized. Although the network that carries out cognitive tasks is distributed, the mental operations that constitute the elements of the task are localized" (p. 18).

The idea expressed here of localized and function-specific nodes encoding or representing mental operations still prevails in the context of many distributed theoretical systems. To a large degree, such theories perpetuate the idea that cognition is represented by specialized and localized mechanisms, although each is a component of a distributed system.

The main point made by all of these function-specific distributed system models is that no single region of the brain is solely responsible for any cognitive process or subprocess. Instead, each of these theories of cognitive brain relationships incorporates the idea of a system of distributed nodes that heavily interact with each other. Identifying these specific functions of these nodes is a major goal of much brain imaging research these days. Unfortunately, as we see in chapter 3, the empirical evidence does not support this kind of specificity, nor, for that matter, does it support the idea of isolatable functional nodes of any kind.

Nevertheless, investigators continue to attempt to attach specific functions to nodes by various experimental techniques; however, this effort sets a goal that may be contradicted by the data. These networks are all heavily interconnected with multiple feedback channels. As a result, it is rarely possible to determine where an activity is initiated or whether or not a part of the system is carrying out either a necessary or a sufficient function (Hilgetag, O'Neil, & Young, 1996.) Ideally, it would be necessary to hold the activities of all except one node constant and manipulate its inputs to determine its role in the system. However, for many technical and conceptual reasons this is not possible. The most significant of these obstacles to carrying out the ideal experiment is that it is probable, if not very likely, that it is the activity of the entire system, indivisible into subunits, that actually instantiates the cognitive process. Nor is it possible to excise one component from such a system and attempt to observe if it serves any singular function. For such a strategy to work, it would require a degree of independence of the various nodes from each other in the manner referred to as "pure insertion" by Sternberg (1969) and Friston et al. (1996). Pure insertion implies that removal of a part of the system would leave all other parts functioning as they did originally. This is a highly unlikely possibility; without pure insertion, any research using surgery or the standard subtraction methods used in brain imaging in an attempt to isolate the function of the excised region would inevitably lead to inconsistent and unreliable results. Any effort, therefore, to divide the system into functional nodes, either surgically or psychologically, would be difficult if not fruitless.

The empirical evidence presented in this book speaks strongly against this concept of function-specific nodes in the brain. By far the predominant finding is that many brain regions are involved in any cognitive processes; thus, they must serve general rather than specific functions. Nevertheless, our persisting opinions concerning the gross anatomical subdivisions of the brain and of localization still lead many investigators to seek to assign specific functions to these nodes or, in many cases, to parts of them.

The persistent idea that the nodes of a distributed system have specific functions can be distinguished from another kind of distributed theory—one that does not carry the neophrenological localization baggage of the past. This alternative theory is discussed in the next section.

4.2.4 Distribution without Function-Specific Nodes

The function-specific theory discussed in the previous section implicitly makes a strong theoretical statement: activity in a group of localized function-specific locations in the brain is the psychoneural equivalent of cognition. However, the alternative now being considered is that the strict localization and specificity of function may no longer be good models of how the brain works. The empirical facts that different cognitive processes can activate different brain regions and that different brain regions may have different functions in different contexts countervails the idea of function specificity by localized operators.

This brings us to a theoretical approach that eschews both functional specificity and spatial localization. It accepts the fact that distribution is a fact but has a different outlook on what the distributed brain mechanisms are like and what role they play in representing cognitive functions.[14] In an earlier book (Uttal, 2009a), I pointed out the following empirical arguments that support a macroscopic theory of brain organization that is neither function specific nor localized.

Distribution
Chapter 3 of this present book strengthens the argument that brain image responses to even the simplest stimuli evoke widely distributed responses throughout the brain. Indeed, the prototypical result of most current research is an extensive listing or depiction of many brain regions that are activated by whatever cognitive process is under investigation rather than a few locales. The more that data are pooled, from subjects and then from experiments, the more broadly distributed are the cumulative responses shown to be. In this context, classical function-specific localization is no

longer a viable theory simply because localization is empirically denied at the most immediate level of data analysis. Furthermore, there are several other properties of current research findings that strongly support this alternative of distributed activity without function specificity. These include the ones that follow.

Anatomical Interconnectedness
Anatomical studies of the brain now make it clear that the various regions of the brain are heavily interconnected. It is, therefore, increasingly likely that no brain region could operate in isolation. Diffusion tensor imaging of the brain highlights the multiple bands of white matter that connect even the most distant regions of the brain. Isolated (i.e., localized) responses are, therefore, logically implausible.

Multifunctionality
Scattered throughout this present book is abundant evidence that every brain region responds in many different cognitive processes; none has any unique role in any particular cognitive process. Brain regions thus must play multiple roles (i.e., be multifunctional) and cannot be function specific.

Weakly Bounded Nodes
Regions of the brain are neither anatomically nor physiologically precisely demarcated. None of the usual brain anatomy mapping methods (e.g., the Brodmann areas) corresponds exactly to the activation regions reported by investigators. Indeed, what constitutes the extent of an activation is arbitrary depending in large part on the thresholds set by the investigator either with the imaging devices themselves or the statistical methods used to analyze the complex data sets coming from a brain imaging device.

This is not to assert that the brain is completely homogeneous in the sense of "mass action" or "equipotentiality" (Lashley, 1950) but rather to acknowledge that the extent of many of the brain regions that might have been considered to be "nodes" are imprecisely, if at all, defined. Thus, the concept of a "node" itself may be a hypothetical construct without hard meaning in this discussion. What we may be talking about is a softly bounded region of the brain in which the boundaries are indistinct and overlapping and the functions general. What appears to be a node to some may be a broad region of the brain to others.

Methodological Sensitivity

Any hope of finding separable nodes with specific functions depends on consistency across methods. If different methods of analysis produce different boundaries, any putative nodes as well as the boundaries themselves would be of questionable reality. The increasing divergence of the shape and extent of activation regions as one increases the size of the pool of data, a finding typical of meta-analyses, suggests that many brain regions, particularly the regions of the cerebral cortex, may have no unique cognitive functional meaning.

Functional Recovery

The remarkable ability of the brain to recover function after trauma or surgery is another argument against both innate functional specialization and localization. For the purposes of this present discussion, the most important aspect of this ability is that it means that there is no genetic, predetermined necessity for a particular place in the brain to have a particular function. If a part is in need of repair, other portions than the injured one can often take over some functions. Why, then, is it not plausible to consider that whatever associations there are between a particular place on the brain and a particular cognitive function are also not fixed? There is the ever-present alternative of a kind of ad-lib selective adaptation and adjustment over the life span. If so, then a large amount of individual difference among people would be expected—exactly the finding we encountered in chapter 3.

Finally, I sum up this alternative version of a distributed system theory of the relationships between the mind and the brain by noting that current evidence supports the following conclusions:

• The brain operates based on a distributed system: many regions (i.e., nodes) are involved in any cognitive task.

• Each node in such a system has multiple functions that can adapt as needed to satisfy cognitive tasks.

• Distributed nodes of a complex are not fixed in terms of their function. They are general-purpose entities that dynamically adjust to the needs of the system.

• Nor are their spatial extents fixed. The regions overlap and have no clear or permanent boundaries, which may change from situation to situation.

- Indeed, the nodes may not exist in any kind of divisible or separable anatomical sense. They may just be softly bounded regions of maximum activity in an otherwise continuous distribution of activity. We may have to consider them as general-purpose computation-capable regions that may be recruited as necessary to represent some cognitive process rather than as function-specific nodes.
- The distributed macroscopic measures obtained with brain imaging devices are almost certainly not the psychoneural equivalents of mind; they are more likely to be cumulative measures of the activity of the vast underlying neuronal networks. These macroscopic measures may actually obscure rather than illuminate the salient neuronal processes.
- It is possible that the distributed responses are not directly associatable with cognition. They may preserve some residual information about the neuronal net information processing; however, they do not preserve the critical information and thus, in principle, cannot explain how brain activity is transmuted into mental activity.
- Brain image responses can vary from subject to subject, from time to time, and from task to task. Therefore, they may not represent a sufficiently stable database from which generalities can be drawn.
- The response of any particular region in a cognitive process is to contribute whatever general-purpose information-processing functions are needed to execute a process.
- Regions of strictly predetermined function and precise localization are limited to the transmission pathways of the sensory and motor pathways.
- The idea that poorly defined cognitive modules—the cognitive constructs of psychology—map directly onto function-specific, narrowly localized nodes of the brain is not supported by current research.
- Furthermore, the idea that these poorly defined cognitive modules map in any simple way onto the anatomy of the brain is an implicit, but unsupported postulate of modern cognitive neuroscience. Considering our empirical results so far, there is no a priori reason that they should.
- In sum, brain imaging defined regions of interest may not be related to cognition in the manner that is implicitly assumed by many current cognitive neuroscientists.

4.3 Theory in Contemporary Cognitive Neuroscience

Now that we have explored some of the theoretical approaches currently holding sway in cognitive neuroscience, it is appropriate to evaluate the kinds of theory that contemporary researchers are inferring specifically from their empirical brain image studies. To reiterate, in the main and with only a few exceptions, brain-imaging techniques can only tell us where some brain activity that might be related to a cognitive process is occurring; they cannot delve into the microcosm of neuronal interconnections at which the key information processes instantiating cognition are probably being carried out.

Thus, most neural theories of cognitive processes proposed at this time based on brain imaging data are theories of which parts of the brain are activated by which cognitive tasks. Deeply embedded in theories of this kind of mind-brain relation remains the localization postulate—namely, that the mechanisms underlying cognitive processes are to be found in particular places on the brain. Not even the strong empirical evidence for broad distribution and the lack of function specificity that has been increasingly forthcoming in recent years have been sufficiently compelling to tear many cognitive neuroscientists loose from the localization postulate. As we see in the previous section, this dominant postulate of discrete function-specific locales has largely been replaced by equally discrete, but multiple, nodes operating as the components of complex brain systems. However, as also expressed in that discussion, even this idea of systems of multiple function-specific nodes is in need of modification given the current empirical situation. Neither localization nor functional specialization seems to be able to characterize the empirical data.

In their place, a further modification consisting of a system characterized by distributed components with nonspecific function has now emerged. When one peels away the superfluous and extraneous details, what we see is the ubiquitous "where" question being asked in one way or another by the preponderance of current researchers using brain imaging devices. This may be phrased in a number of ways:

- Where in the brain is there consistent activation during a cognitive task?
- Which parts of the brain are the neural substrates of cognition?
- What part (or parts) of the brain are activated during a cognitive task?

- What are the relative amounts of activity in two (or more) brain regions during a cognitive task?
- Where are concordant and consistent responses found in the brain during a cognitive task?
- Do all of the parts of an anatomically defined structure of the brain perform in the same manner during a cognitive task?
- Do two (or more) related cognitive tasks produce activity in the same or different areas of the brain?
- Are two cognitive tasks that produce responses in the same area of the brain actually manifestations of the same task?
- Do different categories of objects or individual objects (e.g., tools, animals) activate the same or different areas of the brain?
- In general (acknowledging that cognitive processes are most likely encoded in the brain by distributed systems ofneurons), what are the components or nodes of the system involved in a particular cognitive process?

All of these questions, regardless of the detailed concern of each investigator, are currently being answered predominantly in terms of spatial location and function specificity. This research, therefore, implicitly assumes that there is some kind of regional functional specialization.

It is instructive at this point, to let researchers in this field speak for themselves. After carrying out their research, what conclusions do they draw from their research? These are some of the types of answers emerging from the type of questions just posed. A sampling of these conclusions, all taken verbatim from their articles, now follow.

Taken together, our meta-analysis reveals that animals and tools are categorically represented in visual areas but show convergence in higher-order associative areas in the temporal and frontal lobes in regions that are typically regarded as being involved in memory and/or semantic processing. Our results also reveal that naming tools not only engages visual areas in the ventral stream but also a fronto-parietal network associated with tool use. (Chouinard & Goodale, 2010, p. 409)

The results support distinct dorsal-ventral locations for phonological and semantic processes within the LIFG [left Inferior frontal gyrus]. (Costafreda et al., 2006, p. 799)

The results indicate, unlike results usually reported for adults, children primarily engage the frontal cortex when solving numerical tasks. With age, there may be a shift from reliance on the frontal cortex to reliance on the parietal cortex. In contrast, the frontal parietal and occipito-temporal regions at work during reading are very similar to those reported in adults. (Houde, Rossi, Lubin, & Joliot, 2010, p. 876)

Macroscopic Theories of the Mind-Brain   161

These results support a hypothesized dysfunction of the ACcd [anterior cingulate cognitive division] in ADHD. (Bush et al., 1999, p. 1542)

The WCST [Wisconsin card sorting task] was associated with extensive bilateral clusters of reliable cross-study activity in the lateral prefrontal cortex, anterior cingulate cortex, and inferior parietal lobule. Task switching revealed a similar, although less robust, fronto-parietal pattern with additional clusters of activity in the opercular region of the ventral prefrontal cortex, bilaterally. Response-suppression tasks, represented by studies of the Go/No-go paradigm, showed a large and highly right-lateralized region of activity in the right prefrontal cortex. The activation patterns are interpreted as reflecting a neural fractionation of the cognitive components that must be integrated during the performance of the WCST. (Buchsbaum, Greer, Chang, & Berman, 2005, p. 35)

Analyses of material type showed the expected dorsal-ventral dissociation between spatial and nonspatial storage in the posterior cortex but not in the frontal cortex. Some support was found for left frontal dominance in verbal WM [working memory], but only for tasks with low executive demand. Executive demand increased right lateralization in the frontal cortex for spatial WM. Tasks requiring executive processing generally produce more dorsal frontal activations than do storage-only tasks, but not all executive processes show this pattern. Brodmann's areas (BAs) 6, 8, and 9, in the superior frontal cortex, respond most when WM must be continuously updated and when memory for temporal order must be maintained. (Wager & Smith, 2003, p. 255)

In this review of 100 fMRI studies of speech comprehension and production, published in 2009, activation is reported for prelexical speech perception in bilateral superior temporal gyri; meaningful speech in middle and inferior temporal cortex; semantic retrieval in the left angular gyrus and pars orbitalis; and sentence comprehension in bilateral superior temporal sulci. For incomprehensible sentences, activation increases in four inferior frontal regions, posterior planum temporale, and ventral supramarginal gyrus. These effects are associated with the use of prior knowledge of semantic associations, word sequences, and articulation that predict the content of the sentence. Speech production activates the same set of regions as speech comprehension. In addition, activation is reported for word retrieval in left middle frontal cortex; articulatory planning in the left anterior insula; the initiation and execution of speech in left putamen, pre-SMA, SMA, and motor cortex; and for suppressing unintended responses in the anterior cingulate and bilateral head of caudate nuclei. (Price, 2010, p. 62)

In summary, this meta-analysis found no significant sex difference in functional language lateralization in a large sample of 377 men and 442 women. (Sommer, Aleman, Bouma, & Kahn, 2004, p. 1845)

Inspection of this list of conclusions makes it clear that the spatial locations of the activated brain regions associated with particular cognitive processes are the dominant concern of all of these and most other studies using brain imaging techniques. In a few cases, there is concern with such ancillary issues as the interactions of these spatial entities, or how locations might change over the life span. Nevertheless, the major issue dominating this investigative approach is which part or parts of the brain become activated by a cognitive task. It should be noted, however, that even this simple paradigm results in considerable variability in the answers to the "where" question; the findings of many localization studies are often contraindicated by one investigator or another concerned with the same or a similar problem.

A major methodological issue remaining, therefore, is: Is this (i.e., the search for localized, function-specific nodes) the proper approach? Does it (or can it) lead to a plausible theory of how the brain produces the mind or even suggest new approaches? Alternatively, is this location-oriented emphasis imposed on us by the nature of imaging techniques actually the basis of a bad question? As we have just seen, most current theories are phrased within a very limited context—the empirical findings from their research provide tentative answers to which brain regions may be involved in various cognitive processes, and, for the most part, that is that.[15]

Whether or not this emphasis on localization and function specificity is a productive approach to the actual organizational plan of the brain seems to be a question rarely confronted. Spatial localization is, of course, not without merit—if it is the plan on which the brain is built. There are innumerable practical applications of localization research, and some of the classic conundrums of traditional physiological psychology may be informed to some degree by such research.

However, the preponderance of empirical data currently available to us argues that the concept of the brain as a system of localized function-specific nodes is probably incorrect. Such a perspective ignores that the brain, in empirical fact, functions in a much more distributed and general-purpose mode than is suggested by the standard model. The very fact of conceptualizing the brain as a system of quasi-independent nodes and then applying a method for identifying those nodes may beg the question of how it is actually organized. If we think of the brain as organized in terms of a more or less rigid localization, then we are likely to find it, no matter how artificial may be our view of the nature of those regions. If, on the other hand, we approach the problem from a different theoretical perspective, one emphasizing the distributed, multifunctional, and adaptive nature of brain mechanisms, and a slightly different methodology (one that does not force localized nodes on us), an entirely different theoretical model may emerge.

This is not a new problem; psychology, for many years and for many of the same reasons plaguing current brain imaging studies, has been forcing its observations into a similar mold—cognitive modules. Our cognitive hypothetical constructs and faculties are conceived of as isolatable modules exemplified by such terms as "decision processes," "anxiety," "consciousness," "love," "procedural learning," "attention," and so on. Attempts are made to explore the properties of these isolated modules as if they functioned as independent entities. An alternative view is emerging, however, in which cognition is viewed as an equally integrated and highly complex system in which attempts to isolate one factor or faculty come at great risk. It is possible that it is because of the synergy between the questionable modular hypothesis of cognitive modules and the neurophysiological localization approach imposed us by our current armamentarium of research tools that our data are heavily biased from the cognitive side as well as the neurophysiological side. The old model of localized, function-specific nodes was hard to reject; the new one of distributed, nonspecific nodes may be due for reconsideration.

4.4 Special Issues

To understand why our current emphasis on the location of function-specific nodes may be misdirected, we have to consider several important questions. First, are the empirical data robust; that is, do reliable brain images exist that can be consistently associated with cognitive processes? Second, what is the true nature of the relationship between brain images and cognitive processes? Do brain images represent, encode, or instantiate mental processes in a way that transcends the simple noncausal correlations acceptable if we are only looking for a biomarker? In other words, are the macroscopic neural responses provided by brain imaging devices sufficiently compelling to permit us to assert with confidence that they are the causal neural equivalents whose action *is* cognition? Conversely, if they are not the equivalents, what are they?

It is also important to also ask whether this enormous recent effort in seeking brain image correlates of cognitive processes informs psychological science? Can we resolve conflicts between alternative psychological theories with neural data? Finally, what is it that cognitive neuroscientific theoreticians are really trying to accomplish? To answer this latter question we must delve into the nature of necessary and sufficient conditions for the validity of a theory. The remainder of this section deals with issues such as these.

4.4.1 Can We Trust Brain-Image Correlations with Cognitive Processes?

Let us begin by considering the first of these questions by briefly reviewing the nature of empirical data forthcoming from brain imaging experiments. There are several matters of concern that should be resolved before we attach any deep meaning to them. First, our review of the empirical literature provides compelling support for the conclusion that there is an enormous amount of variability and inconsistency in the findings observed at all levels of analysis. Chapter 3 documents this basic fact for individual subjects, for comparable experiments, and, most consequentially, for meta-analyses, a level for which the pooling of data (and presumably the smoothing) is the most comprehensive. At the present time, only a few investigators (especially see figures 3.1 through 3.3) are explicitly studying variability. Those few who do tackle this critical problem are beginning to question the reliability and, therefore, the validity, of brain imaging data. Increasingly, researchers are noting the lack of replicability at all of these levels. Controversies abound in many studies concerning which brain regions are active during similar, if not identical, cognitive tasks.

Perhaps the best evidence for the shabbiness of the current database is the set of plots of the activation peaks shown in figures 3.4 through 3.9. These figures show the broad dispersion of activation peaks from groups of selected experiments prior to their being pooled into a meta-analysis. Contrary to expectations, the dispersion of responses increases with the number of experiments selected for the meta-analysis. If one were to dissect one these figures and progressively reconstruct it by sequentially plotting the activation peaks from each experiment, one would probably see a progressive increase in the distribution until most of the surface area of the brain would be filled in, as it is the case in these final figures of the accumulation process. The many different sources of bias and variability that plague this field, as discussed in chapter 2, also suggest that inconsistency and variability should be not only ubiquitous in these data but anticipated.

An informative demonstration of the fragile bonds between brain images and cognitive processes, however, can be obtained from the varying patterns of individual brain activations obtained from groups of subjects. There is at least preliminary evidence from the work of Ihnen et al. (2009) that any two groups of subjects, regardless of how they are composed, will produce significant differences in the activation response patterns to any stimulus. If this finding eventually is generalized to parameters other than the gender differences that Ihnen and his colleagues were studying, investigators should expect to be regularly finding some distinctive differences in brain images between any two groups of subjects, no matter how the groups were constructed or for whatever cognitive task is under study. The data on individual differences, furthermore, suggest that such differences between even the best-randomized groups are unavoidable. If so, it would raise serious questions about the validity of many other experiments in which significant brain image differences were reported between two groups of subjects or between experimental and control conditions.

Thus, there appear to be considerable questions about the most basic empirical facts obtained from experiments in which brain regional activations are purported to correlate with cognitive processes. We cannot reject out of hand, therefore, the possibility that the conclusions drawn from the use of this technique are based on a deeply flawed empirical foundation and that we are not really measuring cognitively meaningful brain image differences. Despite the enormous amount of work being done with brain images, it is not inconceivable that much of the resulting data may well be artifacts rather than some correlate or psychoneural mechanism relating the neural and the psychological domains. In other words, we may be reporting something about the nature of "noise" rather than the "signal" in brain imaging experiments.

4.4.2 What Are Brain Images?

If this analysis is correct, then we must confront our next question. Acknowledging the possibility that these neural responses are neither the neural equivalents of thought nor even adequately correlated biomarkers impels us to ask: What, then, are they? The simplest and most direct response to this question is that the fMRI[16] measures the level of oxygen in the blood. The blood oxygen level dependence (BOLD) value, developed by Ogawa et al. (1993), is linked, however, to cognitively significant brain activity by two assumptions. The first assumption is that the BOLD measure reflects local oxygenated blood levels that, in turn, are dependent on cumulative neural activity. The second assumption is that this neural activity is associated with cognitive activity. Collectively, these two assumptions suggest that it is possible to identify cognitively active areas of the brain by subtracting the blood level–related images obtained in control and experimental conditions, respectively.

The first assumption is no longer debatable: it is well documented that the fMRI is measuring blood oxygen levels. The second assumption, however, is far less robust and can be challenged on a number of grounds. The first challenge to the second assumption is the further assumption that the microscopic pattern of underlying neuronal activity reflected in the macroscopic brain image must necessarily be different from one cognitive state to another. However, the BOLD measure is a cumulative measure of the activity of a vast number of microscopic neurons. This means that it is probable for two very different neuronal network states at the microscopic level to produce the same macroscopic BOLD value. When subtracted from each other, therefore, two such brain images could indicate that there was no change between an experimental and a control condition when, in fact, there may have been a vast change in the states of the two underlying neuronal networks.

What has happened, of course, is that in accumulating the microscopic neuronal responses, a vast amount of information has been lost. This means that any null result (no observable difference between brain images obtained in experimental and control conditions or between two groups of subjects) would be indeterminate concerning the state of the underlying neuronal network activity. Thus, it would be an inappropriate indicator of differential cognitive activity because A = B at the macroscopic level is not proof that C and D are identical at the microscopic level. Indeed, this caveat could (and probably does) invalidate the entire subtractive approach to identifying cognitively related areas of the brain or differences in the response of a given area under different stimulus or task conditions.

Given that most cognitive neuroscientists agree that cognition is most likely instantiated in the activity of extraordinarily complex microscopic

Cognitive Activity

→ Distinctive neuronal network activity
→ Increased metabolic activity
→ Deoxygenation of blood
→ Reoxygenation of blood
→ BOLD changes
→ Subtraction of experimental and control findings
→ Distinctive regional activations
→ Supposed macroscopic neural correlates of cognitive activity

Figure 4.1
The logical chain from BOLD to cognition.

neuronal networks, the cumulative brain image measures obtained in an fMRI experiment can thus be logically delinked from cognitive activity. Any observed difference (i.e., an activation) produced by subtraction of two conditions could be due to a multitude of different neural states, some of which might be relevant and some of which might be irrelevant to the cognitive process under consideration. Figure 4.1 (a modification of one presented earlier in Uttal, 2011) summarizes this discussion by highlighting the potential weak links in the chain of logic connecting neuronal network activity through the BOLD levels and thence to cognition.

Potential breaks in this logical chain of contingencies occur at several points. Thus, it should be clear that brain images do not necessarily translate directly to cognitive activity. That is, there is a question of how robust the functional relations are between the observed BOLD-determined brain image and the underlying neuronal network state. In sum, vastly different neuronal network states can produce the same BOLD signal, and the BOLD signal can thus reflect a huge number of alternative neuronal network states. Furthermore, it is not known how (or even if) different neuronal network states associated with a cognitive process are functionally related to different metabolic demands on the brain's metabolism and, thus, to the local blood flow. The point is that both of the key assumptions in using brain imaging techniques to unearth the manner in which the brain encodes cognitive processes are at least subject to some uncertainties.

The second challenge to the second assumption is that vast amounts of information are lost when the microscopic neuronal responses are pooled to produce the cumulative brain responses. As a result, information about the detailed nature of the neuronal network states, the most likely equivalent of cognitive activity, is totally missing from a brain image. Similarly, given the multiplicity of unknown or uncontrollable influences that can be exerted on the BOLD measure, what are believed to be two identical cognitive processes could produce very different BOLD measures, just as two quite different cognitive processes could produce identical BOLD measures.

The third challenge is that the measurements of the blood supply with an fMRI occur on a different time scale than the cognitive processes they are supposed to measure. Thus, the dissimilar timing relations disconnect the thought and the BOLD response.

The fourth challenge, but certainly not the last, is the realization that brain images are noisy and relatively small compared to the background activity in which they are embedded. Therefore, simple subtraction has been replaced by much more elaborate methods to identify more subtly correlated regions of activation. As these powerful techniques have proliferated, the possibility arises that they may be introducing their own properties and artifacts into our interpretation of the results. The point is that the links between blood flow and cognition are not as direct as the present research scene seems prone to accept. Rather, the idea that different regions of the brain have different functions is based on obsolescent assumptions of the simplistic localization theory that pervaded modern cognitive neuroscience until recently. It also depends, as I have just shown, on the uncertainty of several fragile logical steps in the chain of evidence from neuronal activity to cognition.

So the question persists: If the data presented in the form of brain images are not correlates or psychoneural equivalents of cognitive processes, what are they? No one has a full answer to this question, but it is possible that brain images are manifestations of physical or neurochemical processes that have little to do with the neural encoding of cognition per se. Although few would deny that the brain is the organ of mind and all cognitive processes are the product of brain activities, not all brain responses that fMRI systems evoke need necessarily be identifiable with cognitive activities. Furthermore, not all mental activity may be evidenced by observable brain activity, and even what we might call a resting mind may be associated with significant amounts of spontaneous or default brain activity (Fox et al., 2005; Raichle et al., 2001).

An analogy may be drawn between the distributed patterns of the neural activity we call brain images and the pattern of vibration on a tuned metal plate, a device known in physical circles as a Chladni plate.[17] The important point in this analogy is that widespread and very different patterns of oscillation can be produced on a physical surface when it is stimulated in a manner that is resonant with the surface's natural properties. It is by no means certain that this analogy would hold up for the images obtained by an fMRI. However, it seems plausible that for reasons other than the mechanical vibration of a metal plate (perhaps by waves of electrochemical activity), the brain could be producing macroscopic patterns of oscillatory activity. A physical explanation of peaks is that they are the outcomes of nonlinear interactions of waves of neuroelectrical or neurochemical activity that are related to the anatomy and physiology of the brain but not to the encoding of cognitive processes per se.

A similar theory describes the occurrence of "rogue" ocean waves (Onarato, Osborne, & Serio, 2006; Osborne, 2001) where small waves under certain hydrodynamic conditions can superimpose into enormous 100-foot-high monsters. Perhaps, we might speculate, similar nonlinear spatial interactions account for the apparent, but illusory, peak values of brain activations responding to cognitive processes but in a manner that is independent of the stimulus task. In other words, from a cognitive neuroscience perspective, all such responses are irrelevant or noise. Although this is a huge logical leap and is not supported by any current theory or empirical findings, nonlinear processes of this kind could explain why increasingly distributed and inconsistently activated portions of the brain are so often reported.

There are other possibilities to explain the macroscopic patterns of activation on the brain we call brain images, none of which arises to any higher level of plausibility than the ones just mentioned. However, given the fact that the empirical data in this field are so inadequate and that variability is so extensive, the possibility of totally different causal forces should be considered.

Complicating the difficulty of determining what brain images are is the perennial problem of the difference in analytical levels at which brain image data and neuronal net functions are supposed to operate. It may well be that we are simply studying the problem at the wrong level—that is, our attention is being drawn to convenient measurable macroscopic and cumulative phenomena that are actually irrelevant to the problem at hand. The problem, the hard problem, the problem that humankind has long sought to resolve, is: How does the brain make the mind? Correlational studies of the kind to which we are driven by available technology simply may not be able to define the actual causal processes that ideally should be the targets of our investigations.

Because the BOLD values are indeterminate with regard to the state of the underlying neuronal network—the presumptive locus of the information processing that is mind—they may be a risky choice to serve as measures of cognitive activity. What they are designed to do is to answer the "where" question as they define macroscopic regions of the brain putatively associated with cognition. However, given the widespread inconsistency of the data at this macroscopic level, it is not at all clear that they are not just accidental samples of widely distributed and quasi-random activity—accidental samples that can all too easily be extracted from the intractable complexity of the brain.

In sum, we do not have a satisfactory theoretical answer to the question of the relationship between macroscopic brain images and cognitive processes. All of the positive correlations in the world (and it is still problematic whether sufficiently positive correlations do exist) cannot provide the foundation of an explanatory theory of what these signals are or how (or even if) they are causally related to cognitive processes. We should be forewarned, however, by the admonition that if you believe that something exists, you will look for it; if you look for it, you will often find it; especially in a noisy, complex, or quasi-random environment in which everything is possible if not probable.

4.4.3 The Retreat to Methodology

The uncertainty about the relation between brain images and cognitive processes and the relative inconsistency of the empirical database have opened the door to another approach to solving the problem: enhance the measurement methodology to increase detections. The guiding assumption behind this approach is that the responses are so variable and so small given the background activity (i.e., the signal-to-noise ratios are so low) that we must concentrate on developing more sensitive analytic methods that are able to extract these variable and minuscule signals from their much larger camouflaging backgrounds. Logothetis, Pauls, Augath, Trinath, and Oeltermann (2001) sum up this need by noting that:

In all of the measurements, the signal-to-noise ratio of the neural signal was an average of at least one order of magnitude higher than that of the fMRI signals. This observation indicates that the statistical analyses and thresholding methods applied to the haemodynamic responses probably underestimate a great deal of neural activity related to the stimulus or task. . . . (p. 154)

The hope is that by developing ever more powerful statistical methods, cognitive neuroscience would be able to observe subtle and slight differences that would otherwise evade current measurement techniques.

The problem with such an approach, of course, is that the methodology itself comes encumbered with some of its own baggage. One disconcerting possibility, therefore, is that, as the methods become more complex and less transparent, they permit an increasing variety of artifacts to be introduced. (For example, see the discussion of the work of Vul et al., 2009, in chapter 2.) It is also possible for the method to inject its own properties onto the conclusions drawn from an experiment. A classic example of this is the set of "meaningless" solutions to certain kinds of differential equations that are discarded in the development of models of physical systems.[18]

It takes an extraordinary amount of mathematical analysis just to display the results of a single slice or scan of an fMRI. The task is mathematically extremely complicated; our ability just to process the radiofrequency fluctuations obtained with an MRI system into a perceivable pattern was challenging enough to earn Paul Lauterbauer (1929–2007) and Peter Mansfield (b. 1933) the Nobel Prize in 2003.[19]

Closely related is the fact that a huge amount of data, which must be stored for later analysis, is accumulated in any brain imaging experiment. It is important to develop standardized methods so that related data can be accessed and compared for later analysis. An important example of this need is how difficult it is to normalize brain images so that the spatial coordinates of responses both within and between experiments can be compared. This data-comparison task was appreciated early on by Cox and Hyde (1997) but remains a critical problem, especially when we are dealing almost exclusively with "where" questions. Equally complex are the methods for pooling the many images that are collected from a group of subjects or for a group of experiments in a meta-analysis. Similarly, considerable attention has to be given to developing statistical methods in which pooled data from an experimental condition are compared to those from a control condition. It is not always obvious just how different images obtained from two or more experimental conditions are.

Currently, answering questions like these is a major activity of a number of investigators. But are they necessary? Some investigators including Mumford and Nichols (2009) suggested that simple methods (e.g., least squares) are almost as good as more complex and advanced mixed effects models in distinguishing between images. Without effective methods to compare the results from two conditions, the fMRI-based investigator is relegated to the same kind of narrative hand waving so typical in traditional psychological research.

The processing of data from event-related fMRI images is also a continuing activity. In this case, it is not only necessary to extract the individual high-speed images but also to combine them in a consistent manner. Kao, Mandal, Lazar, and Stufken (2009) have suggested a number of different methods for dealing with this special kind of data.

Finally, it must also be remembered that the meta-analysis techniques that form the core of this book are themselves complex methodological tours de force. The work of Turkeltaub et al. (2002) and others discussed in chapter 2 now play central roles in the emerging methodology of brain image analysis.

The point is that in the absence of easily discriminable and quantifiable data, many investigators retreat from the neurophysiological problem itself to stress formal methodologies that are ever more complex in the hope that these techniques will be able to extract reliable signals from extraneous noise. It is possible that some of these powerful techniques will eventually help to unravel some of the neurophysiological mysteries, but it is also possible that this effort is displacement activity assuming that the mysteries of brain representation are solvable and that a signal is truly there when, in empirical fact, neither of these assumptions may be true.

4.4.4 Can Brain Images Resolve Questions Posed by Theoretical Psychology?

Perhaps the most important and contentious theoretical question in the brain image–cognitive neuroscience enterprise deals with whether or not

Macroscopic Theories of the Mind-Brain 173

brain image signals can be used to evaluate what are otherwise purely psychological theories. By a purely psychological theory, I am referring to the models that are invoked to explain behavioral phenomena but that do not involve any explicit neurophysiological assumptions or postulates. A typical theory of this kind is one that invokes hypothetical constructs such as "consolidation," "decision making," "contrast," or "dual processes" among many others but does not specify how these methods might be embodied in neurophysiological mechanisms. Alternatively, these theories might be simple narrations, control theory types of flow charts, or, in a more rigorous version, mathematical or statistical formulations. The key factor that distinguishes a purely psychological theory is that it is devoid of any allusion to neural mechanisms. The question is: Can we distinguish among alternative theories of this kind by alluding to neurophysiological data of any kind, especially fMRI images? In other words, can brain image data inform psychology theory?

To answer this question in any meaningful way depends on an appreciation of the way in which some of the key words are used. In particular, how those of us who have delved into this arcane matter use words such as "theory," "model," "behavior," and "explain" varies from theoretician to theoretician.

First, it is necessary for me to declare my personal philosophical stance with regard to psychology. I am a radical behaviorist. By so categorizing my personal philosophical perspective, I am asserting my basic belief that the prime subject matter of psychology is the publicly observable behavior of the subject. A corollary of this extreme behaviorist postulate is that reductionism of any kind is not possible. That is, our behavior is not reducible in any coherent manner to either cognitive modules or neurophysiological mechanisms. The reasons for this are several, and I have discussed them extensively in an earlier work (for example, Uttal, 1998). In brief summation:

1. The incredible complexity of the neural networks that most likely instantiate cognitive processes precludes any reductive analysis.

2. The barrier due to the impenetrability of the "black box"; that is, our limited ability to measure the inner workings of a closed system for which we have only input-output information.

3. The indeterminate nature of behavioral observations. There are innumerable possible mechanisms that could account for any behavior, and

there is not enough information in the behavior (i.e., they are underdetermined) to discriminate among these possible alternatives.[20]

4. Analogical reasoning cannot produce unique associations between behavior and the underlying neural mechanisms.

5. In fundamental principle, purely psychological theories are descriptive and intrinsically nonreductive. That is, purely psychological theories are, at best, descriptions of observed behavior. In those instances in which there are neurophysiological postulates attached to theories, those neurophysiological postulates are separable from the psychological theories and, in principle, can be tested. However, many other obstacles preclude answers to the question of internal structure.

6. The classic flow diagram (or control system model) of cognitive process is actually a curve-fitting exercise in which the cognitive behavior is approximated by a system of isolated functions and interconnections. Although it is possible to build such a system that may imitate or simulate behavior, there is no convincing evidence that the functions are localized in the brain as they are in models of this kind. For example, lag may be represented by a single, localized system component in the model, but in reality, the mechanisms of lag may be widely distributed throughout the system. It is the existence of properties such as lag that behavior may signal; however, the specific mechanisms cannot be deduced from the system's behavior.

7. The failure of "pure insertion" prohibits researchers from carrying out experiments that parse a complex neural or behavioral system into components that may or may not actually exist.

8. The lack of a good typology of cognitive functions means that any reductive enterprise would always be constrained by the uncertain manner in which cognitive processes are defined.

9. Many of the controversies that attempt to distinguish between two alternative theories are not actually between two alternatives. Instead, many theories, especially verbal ones (but also including mathematical models) are often duals of each other. That is, they may be the same theory hiding under superficial differences that make them appear different when in fact they are operationally identical. Thus, for example, one mathematical theory may seem to differ from another but be directly derivable from the other. Similarly, two verbal theories may seem to be providing different

interpretations of some behavior but, because of the vagueness of vocabulary, be indistinguishable in an operational sense.

10. Finally, the empirical inconsistency of the behavioral data will always make any kind of reductive theory akin to shooting at a moving target.

Given the limits of both neurophysiological and psychological reductionism embedded in this list, radical behaviorists tend to reject the idea that questions about psychological theories can be resolved by neurophysiological data unless the psychological theory has specific and testable neurophysiological postulates or axioms. Indeed, most psychological theories are not reductively explanatory in any sense of the word. They are descriptive or function-fitting exercises taking advantage of the powerful ability of mathematical and statistical tools to plot the course of even very complicated functions. In sum, a purely psychological theory or theoretical question cannot be resolved by allusion to neurophysiological data simply because it does not have neurophysiological postulates to be tested.

The controversy over the plausibility of neurophysiological data informing psychological theory has crystallized in a recent debate between Coltheart (2006), an opponent of the idea that psychological theory controversies can be resolved by brain imaging techniques, and Henson (2006), a proponent of the idea. Both of these works appeared together in a special issue of the journal *Cortex* (2006, volume 42) dealing in large part with this question. The debate was specifically directed at questions concerning what the powerful modern brain imaging techniques can tell us about cognition.

Perhaps the most significant thing that can be said about this debate is that it happened at all. Given the disdain that cognitive neuroscientists typically have toward quasi-philosophical controversies of this kind, it was unusual for the controversy to be made as explicit as it was in the *Cortex* special issue. It is almost axiomatic among laboratory investigators in the field that the brain imaging techniques have opened up an enormous new opportunity to study cognition; the magnitude of the current commitment of resources to cognitive neuroscience is motivated by that assumption. Whether this assumption is justified depends on the outcome of debates such as this one.

Before I begin to discuss the debate itself, it is important to note that both of the two main protagonists went out of their way to be specific about what kind of questions were being debated. Part of the problem in this kind of debate is that the proponents on each side are often not discussing the same issue or assigning the same meaning to their respective vocabularies. In this case, however, both Henson and Coltheart were forthright and helpful in clarifying both what they were considering and what they were not in their discussions. For example, Henson is very specific that in this debate he is supporting the position that that brain-imaging data can be used to "distinguish between competing psychological theories" (p. 94).

Coltheart (2006) very sensibly raised another constraint when he disavowed any argument that would suggest any "in principle" limitation concerning what the future may hold. He argued only that, so far, no one has yet used brain images to tell us anything novel about psychological theory.

Furthermore, both Henson and Coltheart disavowed any interest in the localization problem per se—that is, the determination of which areas of the brain might be associated with cognitive processes. Coltheart, for example, said, "My paper, like Henson's, is concerned solely with the impact of functional neuroimaging on the evaluation of theories that are expressed solely at the psychological level"[21] (p. 323).

The debate was framed in an earlier paper published by Henson (2005) in which he distinguished between two kinds of neurophysiological inference that could be used to distinguish psychological theories—function-to-structure, on the one hand, and structure-to-function, on the other. By function-to-structure inference, he asserted that ". . . a qualitatively different pattern of activity over the brain under two experimental conditions implies at least one different function associated with changes in the independent variable" (p. 193).

By this, I believe he meant that if you can show that there are different brain image responses to two experimental conditions, then you can conclude that there are functional differences (or at least one) between the two conditions. This is a deductive process in which observed brain response differences imply that the psychological processes must be different.

As Henson (2005) pointed out, this does not require the identification of any specific brain regions, only that the brain responses for the two stimulus conditions are different. Thus, a finding of different brain responses implies that they must be instances of different cognitive processes, and, therefore, a functional theory must take into account the "fact" that different cognitive processes are at work. Henson asserted that the only assumption one has to make in this kind of function-to-structure inference is that "the same psychological function will not give rise to different patterns of brain activity within that experiment" (p. 198). In short, according to Henson, brain response differences imply psychological functional differences.

Henson's other kind of inference, structure-to-function, is not a deductive logical process but an inductive one in which the observation of similar or identical brain regional activations implies that similar cognitive functions are at work. The logic here is that if different stimuli produce the same brain responses, there is an increased probability that the two stimuli are part of the same psychological function. In short, he argues that similar brain responses imply similar psychological functions.

Unfortunately, it is more likely that neither of these two means of inferring relationships between neural and psychological responses is true. That is, differences in measured brain responses do not necessarily mean that different cognitive processes are at work; depending on the task, the same brain image data may represent different cognitive tasks, and different cognitive processes may be represented by what are, to the brain imager, the same neural responses. The basis for these assertions is that macroscopic brain images are oblivious to the details of the neuronal network. Henson (2005), nevertheless, went on to give six examples of situations in which he argued these types of inference could reasonably be applied and which he, therefore, offered as evidence for the assumption that brain images could inform psychological theory by distinguishing between alternative theories.

I now turn to Coltheart's response to Henson's argument that brain images can inform psychology theory. Coltheart's strategy was to show that in none of Henson's six examples had this actually occurred. I do not go into the details of these refutations here; in brief, here are the essences of the reasons that Coltheart does not believe that Henson had made his case that neurophysiological data distinguished between alternative purely psychological theories.

Two theories of recognition memory offered by Henson are not different because they both make the same prediction.

Two theories of unexpected memory receive equal support (or equal lack of support) from the relevant neuroimaging studies.

Two theories of the perception of inattended stimuli were not blind to the irrelevant words—an experimental control error.

Two theories of processing facial identity and expression were not actually being tested. Instead, a single theory was accepted as the correct answer, and this experiment, used by Henson as an argument for brain images informing psychological theory, was actually just a search for the brain locales associated with that theory—issues that were explicitly rejected by both Coltheart and Henson.

The same criticism was raised in a facial identity and expression experiment—it was actually a search for the location of relevant brain locations.

Two theories of how we represent others' intentions were equally well supported by the same brain image data.
(Paraphrased from Coltheart, 2006)

Coltheart concluded his essay by challenging his readers to find counterexamples to the original proposition he was arguing, specifically, that no one yet had found a way to distinguish between two alternative theories of psychological function with brain imaging data. The challenge remains open.

At this point, it is useful to bring in some of the other cognitive neuroscientists who participated in the special issue to see if we can resolve the issue. Page (2006), for example, supported Coltheart's position when he also concluded that there had not yet been any evidence that neurophysiological data obtained from brain imaging experiments had been effective in discriminating between cognitive theories. The main point Page made was that purely psychological theories make no testable predictions about the neural underpinnings because there is no contact point between the theory and the imaging data. Finally, he pointed out that the concept of cerebral localization is a major postulate of any attempt to use brain imaging data as a discriminator between alternative theories. The problem he articulated in this context is that localization itself is now increasingly suspect as a model of brain organization. Therefore, it should not be used as a criterion for distinguishing between theories.

To sum up this discussion of this debate, and with the understanding that there are many other investigators who have taken one side or another in it, a balanced view was presented by Poldrack (2006). He pointed out that there is an "epidemic of [inappropriate] reasoning" in cognitive neuroscience. He particularizes this erroneous epidemic as being characterized by "backward" reasoning in which one "reasons from the presence of brain activation to the engagement of a particular cognitive function" (p. 59).

Poldrack's point is that this process is particularly insidious when the connection between a brain region and a cognitive process is weak, as it is in many cognitive neuroscience experiments. He recommends caution in pursuing this kind of backward logic but does not reject the possibility that in some instances it may be useful, particularly, he argues, if approached from a Bayesian analytic point of view.

Should I be able to summon up sufficient chutzpah to score this debate between these serious and competent scholars, I would have to give the win in this debate at this point to Coltheart and those who argue against using brain image imaging data in either creating, evaluating, or discriminating between psychological theories. Although each of these participants in this debate has different reasons for his position, the main core of most of their arguments is the uncertainty of how one builds a logical conceptual bridge between two quite different domains of research. Unless a psychological theory has well-defined neurophysiological postulates, there is nothing, it can be argued, to make such a comparison possible—they simply are not in the same universes of discourse.

In sum, the role of brain imaging findings in informing cognitive theories is currently very limited despite its wide acceptance by the cognitive neuroscience community. Most of the participants in this debate seem to agree that the ultimate goal has not yet been achieved. However, like all good scientists, they also eschew predictions about the immediate and distant future.

4.4.5 On Theoretical Necessity and Sufficiency

Finally, I conclude this chapter on theory in cognitive neuroscience by considering a basic issue in any science. Theories come in many forms. However, among the most important properties of a theory is its intent. Exactly what is the theoretician's intent? Is it to provide a useful and plausible, but not demonstrably unique, description of what is going on? Alternatively, is it to provide the most likely true, definitive, and reductive explanation of some observation that shows how processes at a lower level of analysis lead inexorably to the empirical observation? Is it formulated with the intent of pointing to an otherwise invisible mechanism or of predicting the future course of the system under investigation? These and many other motivations drive investigators interested in some natural phenomenon to develop theories of many different kinds.[22]

One dichotomy that helps us to understand the intent of modern mind-brain theoreticians concerns its sufficiency as opposed to its necessity. Closely related to this dichotomy is whether or not a theory is intended to be descriptive or explanatory in a reductive sense. Let me first make clear what I mean by necessity and sufficiency.

Classical philosophy defined necessity and sufficiency in a direct manner. By necessity, the ancients including Aristotle, Leibniz, and Hume meant "that which has to be true." That is, a necessary theory is one that is required to explain a phenomenon, not just one among many that may possibly be used to describe it. Necessity also implies that a theory is uniquely true; that there are no alternatives unless they are duals of each other. Such a theory is essential to understanding the operation of the system. A necessary theory may allow other supplementary theories, but the necessary one is the one that cannot be done without. Any nonidentical alternatives to such a necessary statement therefore must be incomplete, irrelevant, or inaccurate.

By sufficiency (also known as contingency) is meant that which may not be uniquely true. A sufficient theory may provide a possible and potentially useful representation, but there is a high probability that many other theories will also be capable of describing the same set of phenomena. Of course, both necessary and sufficient theories must incorporate flexibility—the ability to accommodate the flow of new data, of change, and, therefore, that which is necessary at one time might become sufficient at another, and vice versa.

In a more modern context, a sufficient theory is one that is able to describe or explain a body of data in a way that is useful but not unique: it does not exclude alternatives; it only provides one pathway to understanding the behavior of a system such as the mind-brain. That is, any model or theory that fits the data is a priori acceptable as a plausible and sufficient description of those data. However, an unending (literally innumerable according to Moore, 1956) list of other theories may lurk in the background, each of which is also capable of explaining the data to a greater or lesser degree. Only by dissecting the system (if we can) can the multitude of sufficient theories be culled down to a necessary one.

All purely psychological theories are sufficient because of the inaccessibility of lower-level mechanisms. Most current theories of mind-brain relationships are only sufficient because we do not have adequate tests or criteria to distinguish among all plausible alternatives. In such a situation, the best sufficient theory must be distinguished from its competitors by secondary criteria such as simplicity and goodness of fit. Indeed, it is likely that most scientific theories are intended only to be sufficient; most theoreticians appreciate the dynamic nature of our observations as new techniques and points of view evolve; history documents the ebb and flow of sufficient theories as the millennia have passed by.

Narrative psychological theories are sufficient in this sense to the degree that they are plausible or to the degree they can predict behavior. However, it is almost never possible to determine the necessity of theory in psychology. Endless numbers of marginally sufficient narrative models are invoked by psychologists in their efforts to summarize their observations. Unless patently illogical, almost any one of the myriad of theories produced by generations of psychologists may be sufficient. It takes truly nonsensical implications, predictions, or contradictions with other scientific principles to reject a narrative theory. Otherwise, one good "just so story" is as true as another.

The situation with theories based on mathematical or statistical descriptions is also clear—they are at best sufficient. Because formal theories invoke quantitative measures, alternatives can be evaluated by asking such questions as: How well does the mathematical function fit the empirical data or how well does the function predict the future? Another traditional criterion is the answer to the question: Which theory is the simplest? That is, which invokes the smallest number of axioms or postulates or involves the smallest number of derivative steps to arrive at a solution? The simplicity criterion, however, may not work in the case of the mind-brain problem because simplicity in the form of a requirement for the smallest number of axiomatic entities (thank you Mr. Ockham and Mr. Lloyd-Morgan) is not required. The operations of the brain and the mind may not be guided by principles of simplicity but, alternatively, by those of redundancy and stability.

Despite these criteria, mathematical models suffer from the same intrinsic property of sufficiency with which narrative theories must cope. That is, they are not, in principle, unique. There are a multitude of alternate mathematical theories that can compete with one another to provide a plausible model of virtually any process. There is nothing in any mathematical theory that allows us to claim the superiority of one or the other with regard to underlying neural mechanisms beyond those secondary criteria of simplicity and goodness of fit.

A very important point is that no matter how well a mathematical model or theory may fit the data, that model is neutral with regard to the specific nature of underlying mechanisms. That is, we cannot deduce from a mathematical model (or, for that matter, from observable behavior) anything specific about underlying structure. Formal models such as those emerging from statistical and mathematical theories, therefore, can only provide plausible sufficient alternatives.

How then do we distinguish between necessary and sufficient theories? About the only criterion that is available to help us distinguish sufficiency from necessity is some kind of concordance with the findings from other fields of science. For example, no one would suppose that neurons operated with the speed of light given what we know of their biochemistry. Similarly, theories of extrasensory perception are precluded by their conflicts with the laws of standard physics.

This then brings us to the related distinction being made between descriptive and explanatory theories. Descriptive theories are those that by means of words and mathematical expressions describe the course of events in a particular universe of discourse. They are, by nature, classified as sufficient or contingent because descriptive theories are neutral with regard to underlying mechanisms. In principle, there is no way to go from a description to a necessary reductive explanation without examining the mechanics of the system directly.

On the other hand, explanatory theories, to the extent they can be confirmed, are closely associated with necessity. Not only do they describe the behavior of a system but they also explicate how the components of the system or those of a lower level of analysis uniquely lead to the observed behavior. Along with supporting observations, an explanatory theory may be good enough to be considered necessary; however, the criteria for explanation and necessity are very severe.

In cognitive neuroscience, attempts to develop necessary and explanatory theories are rarely achieved. Our ability to develop such a "true" theory is obstructed by many obstacles. Our efforts to build a theory are almost always irretrievably sufficient. For example, neuroreductive models that are based on analogies between the behavior of a neuron and some high-level cognitive process cannot confirm unique associations between the two data domains. Such quasi-theories are based on analogies of the form, shape, or time course of processes but do not have the conceptual or causal links between the two sets of observations to justify classifying them as necessary and explanatory. Of course, there are some theories that seem to be both explanatory and necessary given that we may have enough direct relations between the neural response and the cognitive process to exclude possible alternatives. For example, the coding of wavelength by the retina is no longer talked about in terms of alternative sufficient explanations (e.g., opponent versus trichromatic) but rather in terms of well-established neuroreductive relations (e.g., opponent bipolar neurons and trichromatic receptors) gathered in an empirical way that does not permit any alternative explanations.

Therefore, with the exception of the transmission processes associated with sensory inputs and motor outputs, all of our neuroscientific theories of cognition are at present descriptive and sufficient. Those who propose theoretical explanations of such processes as decision making, much less of such social processes as altruism, are leaping far beyond the available data. What will have to be ascertained in the future is whether these sufficient and descriptive theories will evolve into necessary and reductive ones. If not, we face a future of continued speculation and increasing difficulty in distinguishing between deep understanding and superficial hand waving.

Box 4.1
Chapter Summary

> In sum, theory in cognitive neuroscience, particularly when brain images are used as the neurophysiological indicators, is primarily limited to the specification of where some salient activity might be occurring. The general issue of whether cognitive processes are localized or distributed seems to have been resolved in favor of broad distribution without functional specificity for any region. Brain image data still have not been shown to inform psychology—they are not yet able to help in resolving theoretical conflicts. Most so-called theories are plausible alternatives, sufficient and descriptive but neither necessary nor explanatory.

5 Current Status and Future Needs

5.1 The Issues

Modern cognitive neuroscience is motivated by the search for answers to some fundamental questions—some psychological and some neurophysiological or neuroanatomical. Although the review presented in this book cannot answer any of them with the detail we yearn for, it is possible to see some answers to at least some of them are slowly emerging. The questions include these:

1. Can cognitive processes be divided into quasi-independent functional modules in a manner that allows them to be individually studied?

2. Are cognitive processes executed by specific regions of the brain or by distributed networks of demarcatable nodes?

3. Do demarcatable nodes exist?

4. Are the distributed nodes function specific or general purpose?

5. Are the distributed nodes localized or diffuse?

6. Most germane to the discussion in this book, what is the relation between brain images and cognition? Do brain images correlate with, represent, or encode cognitive processes, or are any observed relations merely illusions emerging from the intractable complexity of a quasi-random system?

7. Are the data obtained in brain imaging experiments reliable and consistent?

8. What are the theoretical implications of brain image and cognitive process comparisons?

5.2 Emerging Principles

I now offer a set of emerging observations and principles based on my personal interpretations that summarize the review carried out in this book. These comments are drawn from comparisons of brain image findings carried out at several levels of analysis—intrasubject, intersubject, between individual experimental reports, and between meta-analyses. This emphasis on empirical comparisons has been chosen because it has the potential to demonstrate more clearly than any other approach the inconsistency and lack of reliability exhibited by current brain imaging studies.

1. The modern development of fMRI and similar systems is one of the most important contributions to medical science in history; it is comparable to such developments as anesthesia and asepsis. Its application to physiological and anatomical investigations and medical diagnosis is profound and indisputable. However, there are increasingly serious doubts about its value in studying the brain mechanisms of cognitive processes.

2. Although our internal and external environments may influence brain activities, it is the activities of the brain that make it the organ of the mind.

3. Although still an unproven hypothesis, the most likely level of analysis at which brain mechanisms become cognition, sentience, and consciousness is the detailed interconnectivity of the huge number of individual neurons of the brain. This is known as the Hebb model. Unfortunately, this level is combinatorially intractable because of its great complexity and the huge number and idiosyncratic nature of its interconnections. Thus, it cannot, for eminently practical reasons, provide answers to the mind-brain problem.

4. With few exceptions (single-neuron studies as the most prominent exception), most recently developed techniques such as the fMRI as well as older ones such as the EEG for studying brain activity are responsive to the cumulative response of millions if not billions of neurons. As a result, they obscure the details of the neuronal interconnections most likely to be associated with cognition. It is still problematical therefore, whether the macroscopic level of observation is actually measuring cognitive processes.

5. Hypothetical cognitive modules are not directly accessible to us. The best of our methods for accessing them—introspection, inference from

behavioral studies, and mathematical models—are all underdetermined. That is, they do not contain enough information to distinguish among innumerable plausible alternative and possible systems.

6. Increasingly often, the results of brain imaging experiments demonstrate that responses are broadly distributed over much of the brain for virtually any cognitive process. Every cognitive process appears to activate many parts of the brain.

7. No region of the brain been found that is uniquely associated with any single cognitive process; every brain region appears to serve multiple functions.

8. There is a long list of potential sources of error, bias, and inconsistency in research using brain images as neurophysiological measures of cognition.

9. A major handicap of all studies of cognitive processes, whether behavioral or neurophysiological, is that psychological science has not developed an operationally useful taxonomy of cognitive processes. It is very difficult to search for the roots of a phenomenon when it cannot be defined with adequate precision.

10. In general, the data obtained from experiments attempting to correlate cognitive processes and brain images are unreliable and inconsistent when replicated (which they rarely are).

11. Intrasubject variability of brain images is relatively large. A single subject given exactly the same task will show major differences from one day to the next.

12. Intersubject variability is even greater that intrasubject variability.

13. Interexperiment variability is also substantial with little convergence on a common answer.

14. Therefore, any effort to develop tools for individual diagnosis from group averages is doomed from the outset.

15. The major exceptions to this observed variability occur in the sensory and motor pathways. Success with these transmission processes has led many investigators to assume incorrectly that a similar level of success is imminent with regard to high-level cognitive processes.

16. In the hope of getting higher experimental power, pooling data from a group of experiments to create a virtual database has become a popular approach. This is what is referred to as a meta-analysis. However, when

data are pooled, much salient information concerning the sources of the original data pool is lost. "Data pooled are data lost!"

17. Meta-analyses of the data from what are assumed to be related groups of experiments typically result in substantial disagreement about which brain areas are activated. This is despite the prevailing assumption that at this level of data pooling, the results should converge. Indeed, differences between meta-analyses are so large as to suggest that the meta-analysis method does not work for cognitive neuroscience studies in the manner originally proposed.

18. It is likely that any two realistically sized groups of subjects, even when randomly selected, will produce some differential brain image activity because of intersubject variability. This can lead to misunderstandings concerning the nature of the controlled differences. Random permutation of the same subject pool, a desirable control, is rarely carried out.

19. When the raw activation regions from a number of supposedly related experiments are plotted on a map of the brain, the typical result is that the entire brain space tends to fill up as more and more experimental findings are introduced. This finding suggests one or both of two possible alternative conclusions: either there are very broad distributions of cognitively generated activity or the responses are not related to the cognitive stimuli in the manner expected.

20. There is little that can be considered to be quantitative "theory" coming from modern brain imaging studies. The general conclusion typically drawn by investigators from this corpus of knowledge is that certain areas of the brain may be associated with particular kinds of cognitive activity. This kind of association appears to be the best that brain imaging can do in the way of theory.

21. In its place, a current metaphor (i.e., a prototheory not yet deserving of the honorific "theory") of cognitively driven brain activity is characterized by the following properties:

- Broadly distributed cognitively related activity
- Broadly interconnected systems
- Multifunctional (i.e., general purpose) system nodes
- Weakly demarcated nodes
- Functional plasticity
- Methodological sensitivity

This is the antithesis of the most widely accepted current metaphor that describes the brain as having:
- Localized cognitively related activity
- Separable system nodes
- Function-specific nodes
- Function-fixed nodes
- Demarcatable nodes
- Methodological insensitivity

22. To a considerable degree, a number of investigators have retreated from dealing with the psychobiology of the problem to the development of methods in the hope that we may be able to extract subtle signals from the background noise.

23. A major issue concerns the ability of brain image data to inform psychology. In general, it appears that it is only in rare and specialized situations that this occurs. In other words, neurophysiological measures such as the fMRI do not seem to be able to suggest new theories or confirm old ones.

24. A related issue is the ability of brain imaging data to resolve theoretical controversies among purely psychological theories with no neurophysiological postulates. Although this is still a contentious issue, it appears that there is not yet any example of a controversy between alternative purely psychological theories having been resolved by neurophysiological measures other than for sensory and motor transmission processes.

25. On the other hand, psychology, as poorly defined as are its constructs, must inform neuroscience. Without adequately precise psychological definitions, for what would cognitive neuroscientists seek?

26. Beyond the somewhat tenuous answers to the "where" question (of regions of interest or of systems of them), little else of diagnostic or theoretical value seems to have yet emerged from the substantial investment in cognitive applications of brain imaging.

27. In general, cognitive neuroscience theories seem to be limited to sufficient (one among many plausible alternatives) and descriptive efforts and have not yet produced necessary (unique) and explanatory (reductive) theories.

28. A critical examination of the vast current database of investigations in which brain imaging systems have been used to determine the location of

brain activity under a variety of cognitive stimuli suggests that many of the findings are method rather than psychobiologically determined. Specifically, because of the complexity of the neuronal networks and the serous loss of salient data by pooling, we may have been reading order into what is actually a complex pattern of quasi-random brain activity.

These concluding statements are based on what we know now; earlier metaphors were based on what we knew a couple of decades ago. No one can predict the future. Those of us who have grown up since World War II can attest to the unpredictability of future history and technology. Who could have predicted the amazing development of such sciences as genetics or of technologies that would permit everyone to carry around his or her own computer and telephone?

5.3   Future Needs

Throughout this book, I have taken a critical and skeptical approach to what I believe is a heavily overblown science—the effort to associate the activity of macroscopic brain images recorded from the brain with cognitive processes. I have pointed to what many cognitive neuroscientists already know: there has been a vast change in our interpretations of what these indicators of electrophysiological activity of the brain mean in just the last decade. The remarkably persistent phrenological and neophrenological points of view of function-specific localization have been largely replaced with ideas of broadly distributed general-purpose systems. How broadly distributed remains a major unknown, but it is possible that a cognitive process may involve nearly the whole brain. Along with this change in the prevailing metaphor, however, has also come a realization that the problem is far more complex than was originally thought. One indication of this complexity is the highly inconsistent nature of and lack of replicability of the empirical data at several levels of examination. Indeed, this inconsistency has led many to challenge the entire enterprise.

There is, therefore, a growing amount of controversy about where the effort to link brain images and cognitive processes should go in the future. It is clear that resolution of these controversies requires specific actions. One way to solve the problem would simply be to throw out the whole business and return to more conventional, if not as promising, lines of

Current Status and Future Needs 191

purely psychological and neurophysiological research. This would hardly be satisfying, however, and would violate all of the standards of centuries of scientific research. Furthermore, the level of controversy itself is a strong argument that this is not a satisfactory outcome; there are too many unknowns, ambiguities, and uncertainties to take that draconian step.

Instead, a more positive approach is required in which we accept the fact that the matter is controversial and suggest some necessary strategies to resolve that controversy. This is a matter that even some of the strongest proponents of the field agree is the right course of action. For example, Poldrack (in press), one of the most articulate and thoughtful supporters of the brain imaging approach to mind-brain studies, has repeatedly pointed out many of the same difficulties I highlighted in my books. He pointed to "fundamental problems in how fMRI has been and continues to be used in cognitive neuroscience" and also noted that ". . . there are fundamental limits to the standard imaging approach that have not been widely appreciated" (p. 1 of preprint).

Poldrack and I, one very optimistic and the other quite conservative about the future, have independently converged on some positive advice to the next generation of cognitive–brain imaging researchers. This advice might help to resolve some of the controversy and determine whether macroscopic brain imaging has been a monumental distraction and fantastically overblown pseudoscience or, to the contrary, actually has opened the door to an ultimate solution to that grand conundrum: How does the brain make the mind?

The following comments are intended to be positive and constructive suggestions for future researchers. Some of them are "hopes" (in Poldrack's words); others are specific admonitions to future researchers to introduce controls that will permit the controversial aspects of this field of cognitive neuroscience to be adjudicated.

Let us deal with Poldrack's suggestions first.

1. Poldrack calls for the improvements of more robust statistical methodologies. He points out that currently applied statistical methods can be misapplied, producing gross errors and artifacts.

2. Then, reflecting the current empirical situation, Poldrack suggests that we move from "blobology to pattern analysis" (p. 2 of preprint). In other words, it is time to ignore the simple concept of function-specific, narrowly

circumscribed spatial localizations and concentrate on uncovering the organization of widely distributed activity, perhaps even at the best achievable pixel level.

3. He then reiterates his long-standing admonition about reverse inference (Poldrack, 2006), that is, inferring cognitive processes from specific brain regional activations.

4. Poldrack's next strategic suggestion is that data should be collected into central stores so that even the most widely separated investigators can pool their raw data, essentially increasing the power of their analyses. This, he also suggests, may permit more powerful central computing facilities to be applied to complex brain image analyses.

5. Finally, Poldrack argues that there is a very profound need for improvement in our "formal descriptions of terms and their relationships" (p. 4 of preprint). That is, psychology must step up to its role of defining cognitive processes in a way that permits both empirical reliability and theoretical progress.

It is with Poldrack's point 5 that I share the most agreement. This was one of the main themes presented in my 2001 book (Uttal, 2001) and still remains one of the most severe handicaps to progress in correlating brain images and cognitive processes. Psychologists with their endless stream of ill-defined cognitive processes, redundant mental faculties, and hypothetical constructs have not yet provided the precise operational definitions necessary to achieve good control over the independent variables of an experiment. The current taxonomy of cognitive processes is totally inadequate. Whether or not the goal of precision in definition can be achieved may be questioned; the need for achieving this goal is not.

It is with this point, therefore, that I now bridge to my own positive suggestions that may also help to point the way to a deeper understanding of what it is that we are accomplishing with the brain imaging approach to cognitive neuroscience. I present these suggestions in the following list. Because a major problem with brain imaging–cognition comparisons is the lack of empirical reliability, the following strategies may help to clarify current ambiguities in the empirical database, reduce the number of spurious results, and determine whether we are dealing with real signals or just noise.

1. Studies such as the ones carried out in this book in which comparisons are made between subjects, experiments, and meta-analyses should become

de rigueur for the field. There has been inadequate consideration of variability at all levels of analysis. As the mathematical and equipment technologies improve, there should be considerable reduction in the noise and convergence on common answers-if they exist. If there is not, serious problems remain for the approach. However, to resolve this problem, attention should be directed at reliability and replicability of comparable studies.

2. Because of the extensive unreliability in the reported literature, publications should be obliged to replicate their findings before acceptance for publication. This replication ideally should require interaction between two independent laboratories keeping the cognitive task as common as possible. Although, this may seem to be an unnecessary addition to the workload of an individual laboratory, the degree of inconsistency as well the large number of effects that turn out to be unsubstantiated suggests that some means of reducing the amount of noise in the publication system is in order. Obligatory replication would act as a filter for marginal papers.

3. Another strategy for improving the quality of research in this field would be to raise the p levels required for statistical significance. Psychology has for years suffered with the problem of disappearing and quickly refuted effects, and it is now increasingly observed in brain imaging studies. As mentioned earlier, $p = .05$ is considered by statisticians and physical scientists to be a ridiculously low criterion for rejecting null hypotheses. It might be well to increase the acceptable criterion in the brain imaging field to $p = .01$. Certainly this would have the effect of many more misses, but it would also serve as a filter for those marginal, unreplicated experiments that clog publications in the field currently.

4. Experimental conditions (especially subject samples) should be permuted to assure that random differences are not producing spurious effects.

5. Meta-analyses, which at first glance seemed so promising, have not turned out to be as powerful as anticipated. Comparable meta-analyses, as we have seen, often substantially disagree. It is not clear what the major problems are among the many sources of bias listed in table 2.1. If they are the influences of the psychobiological variables, this is perfectly understandable—that is what we are trying to understand. However, many of the other sources of bias are methodological and may be obscuring these psychobiological variables. An extensive effort to evaluate and optimize the way meta-analyses are constructed and used is in order.

6. An important mathematical task that also should be pursued is to explore what data pooling means in this context of brain imaging experiments. There are some overly simplistic notions of regional averaging that are uncritically assumed by many among the current crop of investigators. The method of subtracting the results forthcoming from control and experimental conditions to determine the presence or absence of a brain response must be reevaluated.

7. Surprisingly, despite the fact that this form of cognitive neuroscience has been aimed at answering the most fundamental theoretical questions of the relation between mind and brain, there has been relatively little theory forthcoming. It seems to me that the preponderance of effort has purely been an empirical one. The robustness of many of the techniques in common use has been accepted uncritically. Therefore, it seems appropriate to suggest that there is a need for a much more extensive study of the logic of the strategies and technologies. Perhaps this is a characteristic of a new field, and the problem will eventually be overcome by time itself. Nevertheless, a few philosophers mixed in with the neuroscientists might smooth over some of the logical confusions permeating this field.

8. One major problem remains the issue of levels of analysis. Most cognitive neuroscientists accept the Hebbian concept that mind is instantiated at the microscopic level. Although there are compelling reasons why it is so difficult to study these great networks, it seems that a much more concerted effort might be directed at neuronal network theory and particularly at the relation between macroscopic and microscopic findings.

Clearly, there are many tasks for the future, and future developments are unpredictable. It is possible that brain imaging studies of cognition will fulfill the hopes of their proponents. However, if I were to make a bet currently, I would bet that we will never be able to solve the mind-brain problem (or any significant part of it) at the macroscopic levels of analysis now so popular. Not only is the necessary knowledge unobtainable, but it also appears that this is not even the right level of analysis to seek such solutions; therefore, whatever results may be forthcoming from the macroscopic approach may be inconsequential.

The concluding point is that cognitive neuroscience's utilization of brain imaging technology is by no means established as "the" scientific breakthrough it is widely purported to be. Whether it is the extraordinary key to the mind-brain problem is yet to be determined. At present, a necessary task is to continue to determine how successful it will be given what many at the frontier acknowledge to be its conceptual limits and empirical difficulties. It is probably not too much of an exaggeration to claim that evaluating the brain imaging approach to cognitive neuroscience remains one of the most important unresolved challenges of modern science. It is necessary that constructive criticism replace the hyperbole that has characterized much of the field up to now.

An Added Note

A key question asked throughout this book is: Are nodes (i.e., activation peaks) real or merely artifacts of our statistical method? Their reality has just been challenged by the work of Thyreau et al. (in press) and Gonzales-Castillo et al. (2012), who have all argued that the "nodes" are artifacts and that the actual nature of the brain response to a cognitive task is virtually total distribution of brain activity. They account for the appearance of the nodes as (1) inadequate sample sizes, (2) poor signal-to-noise ratios, and (3) biased statistical methods.

Thyreau and his colleagues used a huge data base of 1326 subjects and found ". . . a wide activated pattern, far from being limited to the reasonably expected brain areas, illustrating the difference between statistical significance and practical significance" (p. 1 in preprint).

Similar conclusions were arrived at by Gonzalez-Castillo and his colleagues when they asserted that "We have demonstrated that sparseness of activations in fMRI maps can result from elevated noise levels or overly strict predictive BOLD response models. When noise is sufficiently low and response models versatile enough, activity can be detected with BOLD fMRI in the majority of the brain" (p. 6 in preprint) and ". . . the sparseness of fMRI maps is not a result of localized brain function, but a consequence of high noise and overly strict response models" (p. 1 in preprint).

One interpretation of the data and conclusions of these two groups is that nodes are artifacts. If this is true, then the entire meaning of fMRI responses, the existence of activation nodes, and macroscopic connectionist theories are called into question. If their work is replicated and authenticated, these two contributions may be among the most important articles of the last two decades in brain imaging cognitive neuroscience. If correct, there would be no macroscopic nodes, just evenly distributed responses across the whole brain, and no justification for node-based theories. The whole zeitgeist of recent years of multiple localized activation regions on the brain would have to be reevaluated.

Without "nodes," "regions of interest," and "activation sites" we would have to reevaluate not only almost all of current image-based cognitive neuroscience but the entire commitment to any vestige of the localization theory that has characterized brain sciences for hundreds of years. This is a matter of extreme importance and deserves the equally extreme attention of the research community; replication of this work is vital to legitimate progress in this most important science.

Furthermore, Eklund, Andersson, Josephson, Johannesson, and Knutsson (in press) have shown that reexamination of another very large data base showed that as many as 70% of the datasets they analyzed showed positive results. Based on a .05 criterion, this suggests that there is a proclivity for false positives in much of this data and that any conclusion drawn from this kind of fMRI data is quite possibly invalid.

Notes

Chapter 1

1. This is not to diminish the difficulty of the corresponding problem of comparing time series data such as those obtained with EEGs. My emphasis in this book, however, is on the spatially disparate findings encountered with maps of the kind produced by brain imaging techniques.

2. I have discussed the problems encountered in accepting these two assumptions in greater detail in my earlier books (Uttal, 2001, 2009a, 2011) as well as the progress that has been made in moving away from them.

3. Unfortunately, I cannot find the source of this quotation. I would appreciate anyone calling its source to my attention.

4. Statistical power is defined as the probability that an experiment will correctly reject a false null hypothesis, that is, the probability that the experiment will show that there actually is a difference between two conditions when the difference actually exists. Most important in influencing power is the sample size and its relation to the acceptable confidence level. Although there has been considerable recent interest in requiring a preliminary power analysis of any proposed experiment in order to determine appropriate sample size, this has not routinely been done in the past despite the sage advice of scholars such as Cohen (1992).

5. This is certainly not a new problem for psychology and education. Glass (1976), among others, was instrumental in applying early techniques to the many articles that so often reported inconsistent results in educational research in particular. It is interesting to note that a number of these early meta-analyses concluded that although there were many individual experiments that supported some hypothesis, ultimately the effects disappeared altogether when data were pooled.

6. I use the word "meta-analysis" here in a generic manner much as it is used throughout this literature. In fact, as we see in the next chapter, there is a finer taxonomy of meta-analyses of several different types.

7. There is another form of data pooling that I do not deal with beyond a brief mention in this book. It is called *mega-analysis* and refers to the pooling of the raw results from a group of experiments into a single huge database. Mega-analyses are to be contrasted with meta-analyses in which the already pooled data from a number of experiments are further combined. It should be noted that the results of a meta-analysis and a mega-analysis are not necessarily the same. (See the discussion of the Simpson paradox in chapter 2.)

8. This general statement is partially contradicted by the relative stability of the responses from the early portions of the sensory systems and by the later portions of the motor system. In those regions, there is a far greater degree of stability and far less variability of the recorded brain images than in the areas we traditionally refer to as association areas. Throughout this book my emphasis is on what are now known as the higher cognitive processes.

9. A bad question comes in at least two forms. The first is one that is so loaded with a priori assumptions and postulates that the answer is, in fact, preordained. The second is a question that cannot be answered for reasons of complexity, numerousness, computability, or preexisting proof of intractability.

10. The door left open by Wager and his colleagues is both notable and astute. Not all current investigators are as open minded concerning the possibility that such localizations actually might not exist. Indeed, Wager et al. have passed through this door in what may be a milestone article (Lindquist, Wager, Kober, Bliss-Moreau, & Barrett (in press) concerning the brain basis of emotion. In this paper, they compare a "locationist" approach and a "constructivist" approach. Their analysis suggests that a constructivist model invoking a distributed set of general-purpose, interacting brain regions is activated by a wide variety of emotional as well as nonemotional cognitions.

11. See Loftus (1996) and Nickerson (2000) for insightful discussions of the problems associated with the current use of significance testing and Killeen's (2005) proposed alternative.

12. This subject is treated in detail in what I consider to be one of the most important books in cognitive psychology—Michell's (1999) *Measurement in Psychology; Critical History of a Methodological Concept*. This little-known work should be required reading for everyone who aspires to understand psychological research..

13. Lindquist, Wager, Kober, Bliss-Moreau, and Barrett (in press), in their important review, also deal with the problem of defining the psychological entities used as stimuli in our experiments. They note that the usual method of dealing with modular emotions such as fear, disgust, and happiness is that these are ". . . biologically based, inherited, and cannot be broken down into more basic psychological components" (p. XX). Furthermore, it has also been traditionally assumed that each

of these basic emotions is associated with particular brain locations. However, Lindquist, Wager, and colleagues demonstrate that these psychologically defined modular emotions may actually not be different entities. Instead, they may be better described as the responses of a common neural system. Thus, from a neural point of view, we may have only a single, general property of emotional "excitement" available to us. To that generalized excitement is added a particular quality that depends on the environmental stimulus eliciting the responses. We then particularize these modified responses by adding names such as fear or hate. The conclusion that one might draw from this line of thought is that it is nonsensical to seek distinguishable neural responses to psychologically defined hypothetical constructs.

14. There is no reason to accept .05 as a suitable criterion for cognitive neuroscience experiments. It may be far too low and, therefore, responsible for much of the inconsistency observed in this field. As noted, it surely does not compare to the level of probability required in the physical and engineering sciences.

Chapter 2

1. A classic example of this kind of case study artifact is the association of short-term memory with hippocampal damage in the iconic subject HM by Scoville and Milner (1957). It is rarely mentioned in current discussions, however, that only a portion of their subjects exhibited the archetypical dysfunction—short-term memory failure (three severely; five only moderately; and two not at all). Other than HM, the other subjects in this surgical experiment are generally ignored; nor is the substantial variability of their respective handicaps mentioned. We now know that the nature of the surgical lesions in the sample of 10 subjects was very much more varied than should justify any robust unique association between the hippocampus and short-term memory.

2. Somer and her colleagues also carried out a meta-analysis of 24 studies in addition to the voting procedure. The meta-analysis indicated that there was no difference in the lateralization of language between men and women.

3. The Gaussian weighting function acted as a modifier based on interpeak distances. The further two peaks were apart, the less they mutually reinforced each other in the space. The Gaussian function is only one of many different ways to express this decrease in influence with distance. Many others are possible; in fact, no one knows what the actual distance function is. It is likely to be a much more complex function that differs in three-dimensional direction, that is, anisotropically.

4. For the purposes of the present discussion, the method used by Turkeltaub and his colleagues is our primary focus, not the specific regions they identified as being relevant. To be complete, we should note that the clusters emerging from their

analysis were found in "bilateral motor and superior temporal cortices, pre-supplementary motor area, left fusiform gyrus, and the cerebellum" (p. 765).

5. The relevant mathematical principle is that "every bounded surface has at least one maximum."

6. An important general principle must be reiterated at this point. Although we may observe distributed activity in the brain at the macroscopic level of the brain, these signals do not necessarily represent the "code" or the "psychoneural equivalents" of cognitive processes. Although they may suggest *where* some of the key activity is going on, they tell us nothing about how the brain makes the mind. It is in the microdetails of the actions and interactions of billions of individual neurons at which cognition is most likely to be truly represented.

7. Lipsey and Wilson (2001) as well as Cooper and Hedges build on these skeletons to provide detailed advice on how to carry out meta-analyses. Glass and colleagues (1981), the founders of modern meta-analysis, also give useful advice in their book.

8. This list is a compilation of the ideas and comments made by many authors. The individual items are theirs; the compilation is mine.

9. It is surprising how persistent a current metaphor or model of mind-brain interaction can be in the face of continuing failure to show robust supporting empirical evidence. The EEG and The ERP have been studied for years in the search for correlations between these signals and cognitive processes. In the main, there has been little scientific product or practical application of this extensive effort in cognitive neuroscience. Yet, researchers continue to pursue these wills of the wisp, ignoring the possibility that these signals may be little more than artifacts of random processes of which we still have little knowledge.

10. There has been a persistent tendency to assume that the successes we have had associating neural codes and psychophysical responses in sensory systems (e.g., Hartline & Ratliff, 1957; Hubel & Wiesel, 1959; and DeValois, Abramov, & Jacobs, 1966) mean that comparable success in the search for neural correlates of cognitive processes is going to be promptly forthcoming. In fact, there are such vast differences between the two domains that as much as we know about sensory systems, we may never be able to achieve the same level of understanding of cognitive systems. It cannot be denied that sensory and motor regions of the brain interact with other "association" areas, but this does not mean that they do not preserve such primitive properties as topological consistency.

11. This is not to say that even refereed journals do a perfect job of guarding the gates. Much work of marginal persisting merit regularly fills even the most prestigious journals.

12. A more complete discussion of the relational influences on visual stimuli can be found in my earlier work (Uttal, 1981).

Notes

13. It is interesting to note that even Eysenck had considered the cumulative implications of many studies in a kind of primitive meta-analysis in formulating his attack on psychotherapy.

14. Eysenck went on in discussing Glass's work to state: "If their abandonment of scholarship were to be taken seriously, a daunting but improbable likelihood, it would mark the beginning of a page into the dark age of scientific psychology" (p. 517). In the light of current developments in meta-analyses, Eysenck was, at least momentarily, a bad prophet. We shall have to see what the future holds.

Chapter 3

1. For an enlightening essay on the problem of declining effect size, see the *New Yorker* article by Lehrer (2010) in which he discusses the possibility that many scientific results are essentially random effects that disappear in retests.

2. "Indistinguishably different brain states" does not mean that the brain states are identical. For logical reasons, we must assume that different cognitive processes must be based on different brain states; to do otherwise is to court some kind of dualism. Complexity produces "in practice" limits on distinguishability that cannot be true "in principle."

3. I have to acknowledge that what consistency means in a diagram like this is not easy to quantify. Although there were some regions in this plate that seem to be repeatedly activated, others disappeared and reappeared from day to day with surprising frequency given the attention paid to stabilizing the conditions of the experiment by McGonigle et al. (2000).

4. However, this may be a matter of degree and emphasis. Both groups present evidence for both intra- and intersubject variability, the latter always exceeding the former.

5. Neurosurgeons obviously take this kind of variability into account when they search for functional regions on the human brain during surgery. Their needs are profoundly more important than are those of theoretical cognitive neuroscience, but it is on the evidence from basic research of this kind that they must base their decisions. The evidence for variability in the surgery is epitomized by the "search" procedure used by surgeons to avoid damaging critical brain regions. Even then, following the search it is problematic whether a localized functional region has been found or if a distributed system has been disrupted.

6. Of course, there may be imperceptible clusters that can be teased out of such distributions. In this part of the discussion, however, I am referring to the raw appearance of the data and the dispersed nature of the results of many experiments.

7. It is important to point out that Cabeza and Nyberg did not actually numerically pool any of the voluminous data they examined (that is, they did not carry out a

formal meta-analysis). They simply tabulated the available data, thus providing an easily perceived representation of a complex pattern of findings.

8. The only area that seems to be deficient in activations was the temporal lobe. This was considered by Cabeza and Nyberg to be a technical artifact due to the difficulty prevalent in the period prior to 2000 of recording activations from the temporal lobe.

9. The most notable recent example of the change in interpretation is the work of Lindquist, Wager, Kober, Bliss-Moreau, and Barrett (in press). Their research argues strongly that localization does not work and that the brain responses are far better characterized by broadly distributed systems made up of general-purpose and diffuse functional regions.

10. It would be easy, of course, to set up an electronic display system to produce a true stereoscopic three-dimensional image (see Uttal, 1985, for one such system).

11. There does seem to be a paucity of activation peaks in the frontal and occipital lobes of the brain in this figure. Both findings are surprising because there is a strong visual component to "single-word reading" (which according to current theory should selectively activate the occipital regions) and a strong cognitive component (which is often reputed generally to activate the frontal region). The reasons behind this lacuna in the data are unknown.

12. Stroop (1935) proposed a standard method for evaluating reading speed. He showed that there was an interaction between printed color names and the colors in which the letters were printed. Reading speeds were inhibited if the color and the printed name of the color were not the same. This test has morphed into many different similar "conflict" tests, but it remains one of the most often used cognitive tests because, unlike many other cognitive tasks, it is particularly well defined operationally.

13. My readers are reminded of the important results obtained by Ihnen, Church, Petersen, and Schlaggar (2009). Their results are a compelling argument that almost any two groups of subjects impelled by the force of almost any stimulus and resulting in any form of cognition will produce some differences in a brain imaging study. Their work emphasizes the need for permutation-type experiments as controls in any experiment. This is best done by randomly mixing control and experimental groups to determine if uncontrolled kinds of differences might be producing spurious positive results.

14. This is not to say that the mind-brain problem would have been solved. Even if we were able to answer the macroscopic "where" question with powerful meta-analytical methods, such an answer would not be the same as an explanation of how the brain produces cognitive activity. To answer that question, it is probably necessary to analyze the details of the neuronal connections at the microscopic

level. Unfortunately, as already noted, that may be impossible because of the enormous complexity and irregularity of those connections.

15. The word "concordance" is often used as a synonym for "consistency." It also connotes the idea of similarity or agreement—in this case, the agreement of the activation peaks to cluster in the same place under similar conditions with a high enough probability for the cluster to be considered to be a statistically significant response.

16. The Li et al. report was a comparison of schizophrenic patients and normal controls. Only the data for the normal controls were used in this meta-meta-analysis.

17. It is also of interest to examine Turkeltaub et al.'s table 1 on page 773 of their report in which the inconsistency of the data from the 11 individual PET studies pooled in their meta-analysis is clearly depicted.

18. The Glahn et al. study was carried out with the main goal of comparing normal subjects and schizophrenic patients when they were performing this test. I report here only the results for a normal control group that used the n-back test.

19. The non-Brodmann, narrative responses, for each of the meta-analysis were too numerous to tabulate. Collectively, these meta-analyses implicated virtually all parts of the brain.

20. In an earlier meta-analysis of emotional neuroimaging, Wager, Phan, Liberzon, and Taylor (2003) further complicated the issue by pointing out that they did not find lateralization in the amygdala itself but did find it exhibited for sex differences in nearby regions such as the hippocampus.

Chapter 4

1. Unfortunately, despite the abundance of experimental reports, there have been relatively few comparative brain imaging studies in which reliability has been directly examined. This is a lacuna that I hope this book will at least partially fill.

2. By quasi-random I am referring to a process or system that is so complicated that we cannot perceive the order actually present in it. The process is not formally random; it is just perceptually or computationally disordered to the point that it does not differ operationally from a truly random process. Quasi-randomness is also referred to as being "low discrepancy" and must also be distinguished from pseudorandom. Pseudorandom designates a sequence that is apparently random but that is generated by a deterministic rule. Typically, a pseudorandom sequence is initiated by selecting a "seed" (an arbitrary number) and then applying a formula to generate what cannot be distinguished from a random sequence ex post facto. Such a procedure is useful in cognitive experiments in which sequential dependencies must be controlled.

3. It is also possible that we already have proofs of the intractability of analysis at the microscopic level. See the work of Moore (1956) and Karp (1986) among others. Although no one can predict the future, at the moment it seems clear that there is little hope of explaining cognition or consciousness at the microscopic level.

4. Modern efforts to extend the "organ of the mind" beyond the brain confuse *influence* with *instantiation*. The brain is influenced by both our internal and external environments, but these environmental factors do not encode our thoughts, they only guide them.

5. As fMRI technology has improved, our ability to examine the temporal dynamics of fMRI images has also improved. Indeed, some systems are now able to perform sufficiently rapidly that the time course of the response to a single stimulus can be measured (Boynton, Engel, Glover, & Heeger, 1996) or the average of a number of stimulus-event-triggered responses computed (Friston et al., 1998). Therefore, it is not completely correct to assert that the temporal and magnitude measurements are not possible. However, my concern in this section is with the spatial dimension, the dimension of major interest in current brain imaging science.

6. A few studies have looked at the size of a brain region as a function of a dysfunctional behavioral process, one prominent example being autism (see chapter 3). Although this may seem to represent a counterexample to the generalization that most research can only answer the "where" question, I think it is fair to argue that this is just another way to measure the location and *extent* of the mediating mechanism. In any event, no consistent answer to the relevance of brain component size to autism yet exists.

7. Nothing I say here about the impenetrability of the mind-brain problem should be construed to deny the basic monistic physicalism implicit in modern cognitive neuroscience. The difficulties in answering these questions arise entirely from complexity and technical inaccessibility and not from any cryptodualism.

8. A caveat must be expressed as I begin this discussion. There are in fact four different general classes of mind-brain theories: (1) single cell; (2) neuronal net; (3) field effects; and (4) macroscopic brain regions. My interest in this present book is only with the fourth of these types and its subdivisions, especially as it has been stimulated by current brain imaging techniques. I extensively discussed the first three theoretical approaches in my earlier book, *Neural Theories of Mind* (Uttal, 2005).

9. The main motivation behind the search for biomarkers for these cognitive dysfunctions is that they may represent an "objective" alternative indicator of purely behavioral symptoms. Although no one denies that ADHD and autism are due to abnormalities in brain states, it is likely that the objective foundations of these maladies lie in subtle errors of neuronal interaction that are invisible to the best modern technologies. The probabilities that we will find some biomarker of either of them are probably very low.

10. It should not go unmentioned that the strategy of treating a biomarker *might* be a means of elucidating the actual relationship between the biomarker and the dysfunctional condition. However, that is another story. All that is sought by the biomarker investigator is a sufficiently high correlation to permit diagnosis and, one hopes, prediction.

11. An interesting idea suggested by my colleague John Reich, a social psychologist at Arizona State University, is that very wide distribution of cognitively salient neural activity is supported by the fact that people have only a limited ability to time-share their thoughts. The suggestion is that most, if not all, of our brain resources are involved in any single cognitive process and that it takes that large proportion of our brain's neuronal network to carry out the information processing necessary to instantiate a thought. Thus, because most of the brain is committed to a single thought under this hypothesis, there is not enough extra processing power to think about more than a few things at one time. Of course, this is not the only speculative reason underlying our single-mindedness, but it is an interesting idea.

12. The James-Lange theory argued that emotions were caused by our perceptions of the physiological responses to emotional stimuli. The Canon-Bard theory argued that our emotions caused the physiological responses.

13. Johnson attributes this model to Rick Gilmore without further citation.

14. My first presentation of this "theory" was in Uttal (2011). In that discussion I did not use the word theory because it seemed to me that our models were not specific or substantiated well enough to be dignified by that accolade. I referred to it as a new "metaphor"—a new way of looking at the mind-brain relationship at the macroscopic level. In this book, I have reverted to the use of the word "theory" for consistency with the other topics in this chapter. However, I still believe that current brain models are not so much theories as metaphors that help us to think about the brain. None of them really answers anything about the mind-brain relationship, and the question, "What is the cognitive function of a given brain area?" may be a bad one. Indeed, it is a "point of view" that may be nothing more than a reflection of the inconsistency of the whole corpus of brain imaging studies. The degree of unreliability of the findings in this kind of research suggests that whatever correlations may be observed may be artifacts of our methods rather than neural representations of our thoughts.

15. In some more subtle cases, the spatial pattern may be used as an indicator (for example, when trying to recognize what is being viewed by the subject); however, it is still only the spatial information that is available.

16. In the following discussion, I will be solely concerned with the question as it relates to fMRI brain images. Not only are these the most used method, but many of the criticisms concerning this type of brain images also hold true for PET and other techniques.

17. Ernst Chladni was an eighteenth-century musician and physicist who studied the vibration patterns of a metal plate when it was bowed with a violin bow. Depending on the shape of the plate, where it is bowed, and where it was constrained by being touched, a multiplicity of different patterns of oscillation could be produced. These patterns of oscillation could then be viewed by scattering sand on the surface of the plates. Of course, I am not suggesting there are metal plates in the head; only that some unknown resonant processes independent of or interacting with cognitive processes may be determining the results.

18. This is not to suggest, however, that all models or analyses are worthless. The use of statistical and mathematical methods to describe phenomena has been one of the grandest accomplishments of modern science. However, when findings are ambiguous or inconsistent, there is a powerful incentive to turn to development of the formal methods themselves to see if one can squeeze out more data than would be evident if the signal-to-noise levels were better. In such a situation, it is all too easy to be seduced by circular reasoning in which marginal findings and powerful methods interact to produce false inferences.

19. A notable controversy arose over that year's Nobel Prize. The award ignored the pioneering work of Raymond V. Damadian, who not only held prior patents for the idea but whose laboratory work pioneered the field. In the opinion of many of us, his exclusion was unjustifiable.

20. Unbeknown to most psychologists, this idea was expressed as a proven theorem by an automaton theorist many years ago. Moore (1956) proved that the internal mechanism of a closed system was indeterminate and could not be ascertained from knowledge of its inputs and outputs alone, that is, from its behavior; there were just too many internal possibilities that could produce the same behavior. To simply carry out further research in such a situation does not work because the number of possible alternative mechanisms grows faster than the number of possible experiments distinguishing among them according to Moore.

21. Actually, this was not entirely true. Coltheart subsequently argued that psychological data can be used to discriminate between psychological theories—an inexplicable and inconsistent contradiction to his previously expressed antipathy to the impact of neurophysiological data on psychological theory. I have argued that neither neurophysiological nor behavioral data can discriminate between alternative psychological theories beyond their utility as descriptions. In that case, Ockham's razor and goodness of fit or some such secondary criterion must be invoked to help us to choose among putative theories.

22. The word *theory* has a pejorative connotation that I find both unreasonable and objectionable. "Theory" is sometimes used in the context of mere conjecture, hypothesis, or speculation. I reject these terms as gross misunderstandings of what the word actually means. A theory to me, as noted in the beginning of this book,

is a synoptic summary of an available corpus of scientific data. The meaning of the word *synopsis* in this context is captured by the dictionary definition, "constituting a general view of the whole of a subject." That is, a theory is the inductive or deductive outcome of a grand summary of related individual experiments and observations. Theories come in many forms, but they all seek to provide general conclusions drawn from specific data. A theory is a distillation and smoothing of the results of many experiments. As such, a theory is far more true or accurate and less speculative than the empirical data itself.

Bibliography

Aguirre, G. K., Zarahn, E., & D'Esposito, M. (1998). The variability of human, BOLD hemodynamic responses. *NeuroImage, 8*(4), 360–369.

Anonymous. (1995). A randomized factorial trial assessing early oral captopril, oral mononitrate, and intravenous magnesium sulphate in 85050 patients with suspected acute myocardial infarction. *Lancet, 345*,669–685.

Anonymous. (2003). *National Academy of Sciences. The polygraph and lie detection: A report by the Committee to Review the Scientific Evidence on the Polygraph*. Washington, DC: National Research Council of the United States.

Anonymous. (2004, October). Drugs vs. talk therapy. *Consumer Reports*, 22–29.

Aron, A. R., Gluck, M. A., & Poldrack, R. A. (2006). Long-term test-retest reliability of functional MRI in a classification learning task. *NeuroImage, 29*(3), 1000–1006.

Astin, A. W., & Ross, S. (1960). Glutamic acid and human intelligence. *Psychological Bulletin, 57*, 429–434.

Baas, D., Aleman, A., & Kahn, R. S. (2004). Lateralization of amygdala activation: A systematic review of functional neuroimaging studies. *Brain Research. Brain Research Reviews, 45*(2), 96–103.

Bangert-Drowns, R. L., & Rudner, L. M. (1991). Meta-analysis in educational research. *Practical Assessment, Research & Evaluation, 2*, 1–3.

Bennett, C. M., & Miller, M. B. (2010). How reliable are the results from functional magnetic resonance imaging? *Year in Cognitive Neuroscience, 1191*, 133–155.

Bond, C. F., & Titus, L. J. (1983). Social facilitation—a meta-analysis of 241 studies. *Psychological Bulletin, 94*(2), 265–292.

Borenstein, M., Hedges, L. V., Higgins, J. P. T., & Rothstein, H. R. (2009). *Introduction to meta-analysis*. Hoboken, NJ: John Wiley & Sons.

Bornstein, R. F. (1989). Exposure and affect: Overview and meta-analysis of research. *Psychological Bulletin, 106*, 265–289.

Boynton, G. M., Engel, S. A., Glover, G. H., & Heeger, D. J. (1996). Linear systems analysis of functional magnetic resonance imaging in human V1. *Journal of Neuroscience, 16,* 4207–4221.

Bradstreet, J. J., Smith, S., Baral, M., & Rossignol, D. A. (2010). Biomarker-guided interventions of clinically relevant conditions associated with autism spectrum disorders and attention deficit hyperactivity disorder. *Alternative Medicine Review, 15*(1), 15–32.

Brambilla, P., Hardan, A., Ucelli di Nemi, S., Perez, J., Soares, J. C., & Barale, F. (2003). Brain anatomy and development in autism: Review of structural MRI studies. *Brain Research Bulletin, 61*(6), 557–569.

Broca, P. (1861). Remarques sur la siege de la faculte du language articule suives d'un observation d'aphemie (perte de la parole). *Billons de la Société Anatomique de Paris, 36,* 330–357.

Brodmann, K. (1908). *Vergleichende Lokalisationslehre der Grosshinrnrind.* Lepzig: Barth.

Brown, S., Ingham, R. J., Ingham, J. C., Laird, A. R., & Fox, P. T. (2005). Stuttered and fluent speech production: An ALE meta-analysis of functional neuroimaging studies. *Human Brain Mapping, 25*(1), 105–117.

Brown, S., & Okun, M. A. (in press). Using the caregiver system model to explain the resilience-related benefits older adults derive from volunteering. In M Kent, M. Davis, & J. Reich (Eds). *The resilience handbook: Approaches to stress and trauma.* New York: Routledge.

Buchsbaum, B. R., Greer, S., Chang, W. L., & Berman, K. F. (2005). Meta-analysis of neuroimaging studies of the Wisconsin card-sorting task and component processes. *Human Brain Mapping, 25*(1), 35–45.

Bush, G., Frazier, J. A., Rauch, S. L., Seidman, L. J., Whalen, P. J., Jenike, M. A., et al. (1999). Anterior cingulate cortex dysfunction in attention-deficit/hyperactivity disorder revealed by fMRI and the counting Stroop. *Biological Psychiatry, 45*(12), 1542–1552.

Bush, G., Shin, L. M., Holmes, J., Rosen, B. R., & Vogt, B. A. (2003). The multi-source interference task: Validation study with fMRI in individual subjects. *Molecular Psychiatry, 8*(1), 60–70.

Cabeza, R., & Nyberg, L. (2000). Imaging cognition II: An empirical review of 275 PET and fMRI studies. *Journal of Cognitive Neuroscience, 12*(1), 1–47.

Chein, J. M., Fissell, K., Jacobs, S., & Fiez, J. A. (2002). Functional heterogeneity within Broca's area during verbal working memory. *Physiology & Behavior, 77*(4–5), 635–639.

Chouinard, P. A., & Goodale, M. A. (2010). Category-specific neural processing for naming pictures of animals and naming pictures of tools: An ALE meta-analysis. *Neuropsychologia, 48*(2), 409–418.

Cohen, J. (1992). A power primer. *Psychological Bulletin, 112*(1), 155–159.

Coltheart, M. (2006). What has functional neuroimaging told us about the mind (so far)? *Cortex, 42*(3), 323–331.

Cooper, H. M., & Hedges, L. V. (Eds.). (1994). *The handbook of research synthesis.* New York: Russell Sage Foundation.

Cooper, H. M., Hedges, L. V., & Valentine, J. C. (2009). *The handbook of research synthesis and meta-analysis.* New York: Russell Sage.

Costafreda, S. G. (2009). Pooling FMRI data: Meta-analysis, mega-analysis and multicenter studies. *Frontiers in Neuroinformatics, 3*, 33.

Costafreda, S. G., Fu, C. H. Y., Lee, L., Everitt, B., Brammer, M. J., & David, A. S. (2006). A systematic review and quantitative appraisal of fMRI studies of verbal fluency: Role of the left inferior frontal gyrus. *Human Brain Mapping, 27*(10), 799–810.

Courchesne, E., Pierce, K., Schumann, C. M., Redcay, E., Buckwalter, J. A., Kennedy, D. P., et al. (2007). Mapping early brain development in autism. *Neuron, 56*(2), 399–413.

Cox, R. W., & Hyde, J. S. (1997). Software tools for analysis and visualization of fMRI data. *NMR in Biomedicine, 10*(4–5), 171–178.

Cox, D. R., & Smith, W. L. (1953). The superposition of several strictly periodic sequences of events. *Biometrika, 40*(1–2), 1–11.

Cox, D. R., & Smith, W. L. (1954). On the superposition of renewal processes. *Biometrika, 41*(1–2), 91–99.

Critchley, M., & Critchley, E. A. (1998). *John Hughlings Jackson: Father of English neurology.* New York: Oxford University Press.

D'Esposito, M., Deouell, L. Y., & Gazzaley, A. (2003). Alterations in the bold FMRI signal with ageing and disease: A challenge for neuroimaging. *Nature Reviews Neuroscience, 4*(11), 863–872.

Devalois, R. L., Abramov, I., & Jacobs, G. H. (1966). Analysis of response patterns of LGN cells. *Journal of the Optical Society of America, 56*(7), 966.

Dickersin, K. (2005). Publication bias: Recognizing the problem, understanding its origins and scope, and preventing harm. In H. R. Rothstein, A. J. Suton, & M. Borenstein (Eds.), *Publication bias in meta-analysis—prevention, assessment, and adjustments* (pp. 9–33). New York: John Wiley & Sons.

Dickstein, S. G., Bannon, K., Castellanos, F. X., & Milham, M. P. (2006). The neural correlates of attention deficit hyperactivity disorder: An ALE meta-analysis. *Journal of Child Psychology and Psychiatry, and Allied Disciplines, 47*(10), 1051–1062.

Donovan, C. L., & Miller, M. B. (2008, December 1–4). *An investgation of individual variability in brain activity during episodic encoding and retrieval.* Paper presented at the 26th Army Science Conference, Orlando Florida.

Duncan, K. J., Pattamadilok, C., Knierim, I., & Devlin, J. T. (2009). Consistency and variability in functional localisers. *NeuroImage, 46*(4), 1018–1026.

Dunkin, M. J., & Biddle, B. J. (1974). *The study of teaching.* New York: Holt.

Eklund, A., Andersson, M., Jasephson, C., Johannesson, M., and Knutsson, H. (in press). Does parametric fMRI analysis with SPM yield valid results? An empirical study of 1484 rest data states. *Neuroimages.*

Eden, G. F., Jones, K. M., Cappell, K., Gareau, L., Wood, F. B., Zeffiro, T. A., et al. (2004). Neural changes following remediation in adult developmental dyslexia. *Neuron, 44*(3), 411–422.

Elliott, H. C. (1907). *Textbook of neuroanatomy.* Philadelphia: Lippincott.

Ellison-Wright, I., Ellison-Wright, Z., & Bullmore, E. (2008). Structural brain change in attention deficit hyperactivity disorder identified by meta-analysis. *BMC Psychiatry, 8,* 51.

Evans, A. C., Collins, D. L., & Milner, B. (1992). An MRI-based stereotactic atlas from 250 young normal subjects. *Society of Neuroscience Abstracts, 18,* 408.

Eysenck, H. J. (1952). The effects of psychotherapy: An evaluation. *Journal of Consulting Psychology, 16*(5), 319–324.

Eysenck, H. J. (1978). An exercise in mega-silliness. *American Psychologist, 33*(5), 517.

Eysenck, H. J. (1994). Systematic reviews—metaanalysis and its problems. *British Medical Journal, 309*(6957), 789–792.

Feinstein, A. R. (1995). Metaanalysis—statistical alchemy for the 21st-century. *Journal of Clinical Epidemiology, 48*(1), 71–79.

Fiedler, K. (2011). Voodoo correlations are everywhere—not only in neuroscience. *Perspectives on Psychological Science, 6,* 163–171.

Fisher, R. A. (1932). *Statistical methods for research workers* (4th ed.). Edinburgh: Oliver and Boyd.

Fox, M. D., Snyder, A. Z., Vincent, J. L., Corbetta, M., Van Essen, D. C., & Raichle, M. E. (2005). The human brain is intrinsically organized into dynamic, anticorre-

Bibliography

lated functional networks. *Proceedings of the National Academy of Sciences of the United States of America*, 102(27), 9673–9678.

Fox, P. T., Parsons, L. M., & Lancaster, J. L. (1998). Beyond the single study: Function/location metanalysis in cognitive neuroimaging. *Current Opinion in Neurobiology*, 8(2), 178–187.

Fox, P. T., & Woldorff, M. G. (1994). Integrating human brain maps. *Current Opinion in Neurobiology*, 4(2), 151–156.

Friston, K. J., Fletcher, P., Josephs, O., Holmes, A., Rugg, M. D., & Turner, T. (1998). Event related fMRI: Characterizing differential responses. *NeuroImage*, 7, 30–40.

Friston, K. J., Holmes, A. P., Worsley, K. J., Poline, J. P., Frith, C. D., & Frackowiak, R. S. J. (1995). Statistical parametric maps in functional imaging: A general linear approach. *Human Brain Mapping*, 2, 189–210.

Friston, K. J., Price, C. J., Fletcher, P., Moore, C., Frackowiak, R. S. J., & Dolan, R. J. (1996). The trouble with cognitive subtraction. *NeuroImage*, 4(2), 97–104.

Fusar-Poli, P., Placentino, A., Carletti, F., Allen, P., Landi, P., Abbamonte, M., et al. (2009). Laterality effect on emotional faces processing: ALE meta-analysis of evidence. *Neuroscience Letters*, 452(3), 262–267.

Gall, F. J., & Spurzheim, J. C. (1808). *Recherches sur le système nerveaux en genral, et sur celui du cerveau en particulier*. Paris: Academie de Sciences, Paris Memoires.

Garey, L. J. (Ed.). (1994). *Brodmann's localization in the cerebral cortex*. London: Imperial College Press.

Gilovich, T., Vallone, R., & Tversky, A. (1985). The hot hand in basketball—on the misperception of random sequences. *Cognitive Psychology*, 17(3), 295–314.

Glahn, D. C., Ragland, J. D., Abramoff, A., Barrett, J., Laird, A. R., Bearden, C. E., et al. (2005). Beyond hypofrontality: A quantitative meta-analysis of functional neuroimaging studies of working memory in schizophrenia. *Human Brain Mapping*, 25(1), 60–69.

Glass, G. V. (1976). Primary, secondary, and meta-analysis of research. *Educational Researcher*, 5, 3–8.

Glass, G. V., McGaw, B., & Smith, M. (1981). *Meta-analysis in social research*. Newbury Park, CA: Sage.

Gonzalez-Castillo, J., Saad, Z. S., Handwerker, D. A., Inati, S. J., Brenowitz, N., & Bandetini, P. A. (2012). Whole brain, time-locked activation with simple tasks revealed using massive averaging and model-free analysis. *Proceedings of the National Academy of Sciences*, 109, 5487–5492.

Gould, S. J. (1987). *Urchin in the storm: Essays about books and ideas*. New York: Norton.

Grunling, C., Ligges, M., Huonker, R., Klingert, M., Mentzel, H. J., Rzanny, R., et al. (2004). Dyslexia: The possible benefit of multimodal integration of fMRI- and EEG-data. *Journal of Neural Transmission, 111*(7), 951–969.

Guzzo, R. A., Jette, R. D., & Katzell, R. A. (1985). The effects of psychologically based intervention programs on worker productivity—a meta-analysis. *Personnel Psychology, 38*(2), 275–291.

Hartline, H. K., & Ratliff, F. (1957). Inhibitory interaction of receptor units in the eye of Limulus. *Journal of General Physiology, 40*(3), 357–376.

Hebb, D. O. (1949). *The organization of behavior: A neuropsychological theory.* New York: Wiley.

Henson, R. (2005). What can functional neuroimaging tell the experimental psychologist? *Quarterly Journal of Experimental Psychology Section a: Human. Experimental Psychology, 58*(2), 193–233.

Henson, R. (2006). What has (neuro)psychology told us about the mind (so far)? A reply to Coltheart (2006). *Cortex, 42*(3), 387–392.

Hilgetag, C. C., O'Neill, M. A., & Young, M. P. (1996). Indeterminate organization of the visual system. *Science, 271*(5250), 776–777.

Hogben, J. (2011). A plea for purity. *Australian Journal of Psychology, 48*, 172–177.

Houde, O., Rossi, S., Lubin, A., & Joliot, M. (2010). Mapping numerical processing, reading, and executive functions in the developing brain: An fMRI meta-analysis of 52 studies including 842 children. *Developmental Science, 13*(6), 876–885.

Hubel, D. H., & Wiesel, T. N. (1959). Receptive fields of single neurones in the cats striate cortex. *Journal of Physiology (London), 148*(3), 574–591.

Hunter, J. E., & Schmidt, F. L. (1990). *Methods of meta-analysis: Correcting error and bias in research findings.* Newbury Park, CA: Sage.

Ihnen, S. K. Z., Church, J. A., Petersen, S. E., & Schlaggar, B. L. (2009). Lack of generalizability of sex differences in the fMRI BOLD activity associated with language processing in adults. *NeuroImage, 45*(3), 1020–1032.

James, W. (1890). *The principles of psychology.* New York: H. Holt and Co.

Javitt, D. C., Spencer, K. M., Thaker, G. K., Winterer, G., & Hajos, M. (2008). Neurophysiological biomarkers for drug development in schizophrenia. *Nature Reviews. Drug Discovery, 7*(1), 68–83.

Johnson, M. H. (1995). The development of visual attention. In M. S. Gazzaniga (Ed.), *The cognitive neurosciences* (pp. 735–747). Cambridge, MA: MIT Press.

Kao, M. H., Mandal, A., Lazar, N., & Stufken, J. (2009). Multi-objective optimal experimental designs for event-related fMRI studies. *NeuroImage, 44*(3), 849–856.

Karp, R. M. (1986). Combinatorics, complexity, and randomness. *Communications of the ACM*, *29*(2), 98–109.

Keller, S. S., Highley, J. R., Garcia-Finana, M., Sluming, V., Rezaie, R., & Roberts, N. (2007). Sulcal variability, stereological measurement and asymmetry of Broca's area on MR images. *Journal of Anatomy*, *211*(4), 534–555.

Kemp, S. (1996). *Cognitive psychology in the Middle Ages*. Westport, CT: Greenwood Press.

Killeen, P. R. (2005). An alternative to null-hypothesis significance tests. *Psychological Science*, *16*(5), 345–353.

Kober, H., Barrett, L. F., Joseph, J., Bliss-Moreau, E., Lindquist, K., & Wager, T. D. (2008). Functional grouping and cortical-subcortical interactions in emotion: A meta-analysis of neuroimaging studies. *NeuroImage*, *42*(2), 998–1031.

Kober, H., & Wager, T. D. (2010). Meta-analysis of neuroimaging data. *Brain and Behavior*, *1*, 293–300.

Krain, A. L., Wilson, A. M., Arbuckle, R., Castellanos, F. X., & Milham, M. P. (2006). Distinct neural mechanisms of risk and ambiguity: A meta-analysis of decision-making. *NeuroImage*, *32*(1), 477–484.

Kriegeskorte, N., Simmons, W. K., Bellgowan, P. S. F., & Baker, C. I. (2009). Circular analysis in systems neuroscience: The dangers of double dipping. *Nature Neuroscience*, *12*(5), 535–540.

Kringelbach, M. L., & Rolls, E. T. (2004). The functional neuroanatomy of the human orbitofrontal cortex: Evidence from neuroimaging and neuropsychology. *Progress in Neurobiology*, *72*(5), 341–372.

Lacadie, C. M., Fulbright, R. K., Rajeevan, N., Constable, R. T., & Papademetris, X. (2008). More accurate Talairach coordinates for neuroimaging using non-linear registration. *NeuroImage*, *42*(2), 717–725.

Laird, A. R., Fox, P. M., Price, C. J., Glahn, D. C., Uecker, A. M., Lancaster, J. L., et al. (2005). ALE meta-analysis: Controlling the false discovery rate and performing statistical contrasts. *Human Brain Mapping*, *25*(1), 155–164.

Laird, A. R., McMillan, K. M., Lancaster, J. L., Kochunov, P., Turkeltaub, P. E., Pardo, J. V., et al. (2005). A comparison of label-based review and ALE meta-analysis in the Stroop task. *Human Brain Mapping*, *25*(1), 6–21.

Land, E. H. (1977). Retinex theory of color-vision. *Scientific American*, *237*(6), 108.

Lashley, K. S. (1950). In search of the engram. *Symposia of the Society for Experimental Biology*, *4*, 454–482.

Lehrer, J. (2010, December 13). The truth wears off. *The New Yorker*.

LeLorier, J., Gregoire, G., Benhaddad, A., Lapierre, J., & Derderian, F. (1997). Discrepancies between meta-analyses and subsequent large randomized, controlled trials. *New England Journal of Medicine, 337*(8), 536–542.

Levy, S. E., Mandell, D. S., & Schultz, R. T. (2009). Autism. *Lancet, 374*(9701), 1627–1638.

Li, H. J., Chan, R. C. K., McAlonan, G. M., & Gong, Q. Y. (2010). Facial emotion processing in schizophrenia: A meta-analysis of functional neuroimaging data. *Schizophrenia Bulletin, 36*(5), 1029–1039.

Lieberman, M. D., & Cunningham, W. A. (2009). Type I and type II error concerns in fMRI research: Re-balancing the scale. *Social Cognitive and Affective Neuroscience, 4*(4), 423–428.

Lindquist, K. A., Wager, T. D., Kober, H., Bliss-Moreau, E., & Barrett, L. F. (2012). The brain basis of emotion: A meta-analytic review. *Behavioral and Brain Sciences*, in press.

Lipsey, M. W., & Wilson, D. B. (1993). The efficacy of psychological, educational, and behavioral treatment. Confirmation from meta-analysis. *American Psychologist, 48*(12), 1181–1209.

Lipsey, M. W., & Wilson, D. (2001). *Practical meta-analysis*. New York: Sage.

Loftus, G. R. (1996). Psychology will be a much better science when we change the way we analyze data. *Current Directions in Psychological Science, 5*, 161–171.

Logothetis, N. K., Pauls, J., Augath, M., Trinath, T., & Oeltermann, A. (2001). Neurophysiological investigation of the basis of the fMRI signal. *Nature, 412*, 150–157.

Maccoby, E. E., & Jacklin, C. N. (1974). *The psychology of sex differences*. Stanford, CA: Stanford University Press.

Maldjian, J. A., Laurienti, P. J., Driskill, L., & Burdette, J. H. (2002). Multiple reproducibility indices for evaluation of cognitive functional MR imaging paradigms. *AJNR. American Journal of Neuroradiology, 23*(6), 1030–1037.

McGonigle, D. J., Howseman, A. M., Athwal, B. S., Friston, K. J., Frackowiak, R. S. J., & Holmes, A. P. (2000). Variability in fMRI: An examination of intersession differences. *NeuroImage, 11*(6), 708–734.

Michell, J. (1999). *Measurement in psychology: Critical history of a methodological concept*. Cambridge, UK: Cambridge University Press.

Miller, M. B., Donovan, C. L., Van Horn, J. D., German, E., Sokol-Hessner, P., & Wolford, G. L. (2009). Unique and persistent individual patterns of brain activity across different memory retrieval tasks. *NeuroImage, 48*(3), 625–635.

Miller, M. B., & Van Horn, J. D. (2007). Individual variability in brain activations associated with episodic retrieval: A role for large-scale databases. *International Journal of Psychophysiology*, *63*(2), 205–213.

Miller, M. B., Van Horn, J. D., Wolford, G. L., Handy, T. C., Valsangkar-Smyth, M., Inati, S., et al. (2002). Extensive individual differences in brain activations associated with episodic retrieval are reliable over time. *Journal of Cognitive Neuroscience*, *14*(8), 1200–1214.

Modha, D. S., & Singh, R. (2010). Network architecture of the long-distance pathways in the macaque brain. *Proceedings of the National Academy of Sciences of the United States of America*, *107*(30), 13485–13490.

Moore, E. F. (1956). Gedanken-experiments on sequential machines. In E. F. Shannon & J. McCarthy (Eds.), *Automata studies* (pp. 129–153). Princeton, NJ: Princeton University Press.

Mostofsky, S. H., Schafer, J. G. B., Abrams, M. T., Goldberg, M. C., Flower, A. A., Boyce, A., et al. (2003). fMRI evidence that the neural basis of response inhibition is task-dependent. *Brain Research. Cognitive Brain Research*, *17*(2), 419–430.

Mumford, J. A., & Nichols, T. (2009). Simple group fMRI modeling and inference. *NeuroImage*, *47*, 1469–1475.

Munk, H. (1881). *Uber die Funktionen der Grosshirnrinde*. Berlin: Hirschwald.

Murphy, F. C., Nimmo-Smith, I., & Lawrence, A. D. (2003). Functional neuroanatomy of emotions: A meta-analysis. *Cognitive, Affective & Behavioral Neuroscience*, *3*(3), 207–233.

Nee, D. E., Wager, T. D., & Jonides, J. (2007). Interference resolution: Insights from a meta-analysis of neuroimaging tasks. *Cognitive, Affective & Behavioral Neuroscience*, *7*(1), 1–17.

Neumann, J., Lohmann, G., Derrfuss, J., & von Cramon, D. Y. (2005). Meta-analysis of functional imaging data using replicator dynamics. *Human Brain Mapping*, *25*(1), 165–173.

Newman, S. D., Greco, J. A., & Lee, D. (2009). An fMRI study of the Tower of London: A look at problem structure differences. *Brain Research*, *1286*, 123–132.

Nickerson, R. S. (2000). Null hypothesis significance testing: A review of an old and continuing controversy. *Psychological Methods*, *5*(2), 241–301.

Ogawa, S., Menon, R. S., Tank, D. W., Kim, S. G., Merkle, H., Ellermann, J. M., et al. (1993). Functional brain mapping by blood oxygenation level-dependent contrast magnetic-resonance-imaging—a comparison of signal characteristics with a biophysical model. *Biophysical Journal*, *64*(3), 803–812.

Onarato, M., Osborne, A. R., & Serio, M. (2006). Modulational instability in crossing wave states: A possible mechanism for the formation of freak waves. *Physical Review Letters, 96*(014503), 1–4.

Osborne, A. R. (2001). The random and deterministic dynamics of 'rogue waves' in unidirectional deep-water trains. *Marine Structures, 14,* 275–293.

Owen, A. M., McMillan, K. M., Laird, A. R., & Bullmore, E. (2005). N-back working memory paradigm: A meta-analysis of normative functional neuroimaging. *Human Brain Mapping, 25*(1), 46–59.

Page, M. P. (2006). What can't functional neuroimaging tell the cognitive psychologist? *Cortex, 42*(3), 428–443.

Paloyelis, Y., Mehta, M. A., Kuntsi, J., & Asherson, P. (2007). Functional MRI in ADHD: A systematic literature review. *Expert Review of Neurotherapeutics, 7*(10), 1337–1356.

Papez, J. W. (1937). A proposed mechanism of emotion. *Archives of Neurology and Psychiatry, 38*(4), 725–743.

Paus, T. (1996). Location and function of the human frontal eye-field: A selective review. *Neuropsychologia, 34*(6), 475–483.

Phan, K. L., Wager, T., Taylor, S. F., & Liberzon, I. (2002). Functional neuroanatomy of emotion: A meta-analysis of emotion activation studies in PET and fMRI. *NeuroImage, 16*(2), 331–348.

Picard, N., & Strick, P. L. (1996). Motor areas of the medial wall: A review of their location and functional activation. *Cerebral Cortex, 6*(3), 342–353.

Poldrack, R. A. (2006). Can cognitive processes be inferred from neuroimaging data? *Trends in Cognitive Sciences, 10*(2), 59–63.

Poldrack, R. A. (in press). The future of fMRI in cognitive neuroscience. *NeuroImage*.

Posner, M. I., & Rothbart, M. K. (2007). Research on attention networks as a model for the integration of psychological science. *Annual Review of Psychology, 58,* 1–23.

Posner, M. I., Petersen, S. E., Fox, P. T., & Raichle, M. E. (1988). Localization of cognitive operations in the human brain. *Science, 240*(4859), 1627–1631.

Price, C. J. (2010). The anatomy of language: A review of 100 fMRI studies published in 2009. *Year in Cognitive Neuroscience, 1191,* 62–88.

Raemaekers, M., Vink, M., Zandbelt, B., van Wezel, R. J. A., Kahn, R. S., & Ramsey, N. F. (2007). Test-retest reliability of fMRI activation during prosaccades and antisaccades. *NeuroImage, 36*(3), 532–542.

Raichle, M. E., MacLeod, A. M., Snyder, A. Z., Powers, W. J., Gusnard, D. A., & Shulman, G. L. (2001). A default mode of brain function. *Proceedings of the National Academy of Sciences of the United States of America, 98*(2), 676–682.

Rau, S., Fesl, G., Bruhns, P., Havel, P., Braun, B., Tonn, J. C., et al. (2007). Reproducibility of activations in Broca area with two language tasks: A functional MR imaging study. *AJNR. American Journal of Neuroradiology*, *28*(7), 1346–1353.

Richman, D. P., Stewart, R. M., Hutchinson, J. W., & Caviness, V. S. (1975). Mechanical model of brain convolutional development. *Science*, *189*(4196), 18–21.

Robb, R. A., & Hanson, D. P. (1991). A software system for interactive and quantitative visualization of multidimensional biomedical images. *Australasian Physical & Engineering Sciences in Medicine*, *14*(1), 9–30.

Rosenthal, R. (1979). The "file drawer problem" and tolerance for null results. *Psychological Bulletin*, *86*, 638–641.

Rosenthal, R., & DiMatteo, M. R. (2001). Meta-analysis: Recent developments in quantitative methods for literature reviews. *Annual Review of Psychology*, *52*, 59–82.

Ross, E. D. (2010). Cerebral localization of functions and the neurology of language: Fact versus fiction or is it something else? *Neuroscientist*, *16*(3), 222–243.

Schramm, W. (1962). Learning from instructional television. *Review of Educational Research*, *32*(2), 156–167.

Scoville, W. B., & Milner, B. (1957). Loss of recent memory after bilateral hippocampal lesions. *Journal of Neurology, Neurosurgery, and Psychiatry*, *20*(1), 11–21.

Sears, D. O. (1986). College sophomores in the laboratory—influences of a narrow database on social-psychology view of human-nature. *Journal of Personality and Social Psychology*, *51*(3), 515–530.

Sergerie, K., Chochol, C., & Armony, J. L. (2008). The role of the amygdala in emotional processing: A quantitative meta-analysis of functional neuroimaging studies. *Neuroscience and Biobehavioral Reviews*, *32*(4), 811–830.

Shapiro, D. A., & Shapiro, D. (1982). Meta-analysis of comparative therapy outcome studies—a replication and refinement. *Psychological Bulletin*, *92*(3), 581–604.

Shapiro, S. (1994). Metaanalysis shmeta-analysis. *American Journal of Epidemiology*, *140*(9), 771–778.

Shercliffe, R. J., Stahl, W., & Tuttle, M. P. (2009). The use of meta-analysis in psychology: A superior vintage or the casting of old wine in new bottles? *Theory & Psychology*, *19*(3), 413–430.

Sim, I., & Hlatky, M. A. (1996). Growing pains of meta-analysis—advances in methodology will not remove the need for well designed trials. *British Medical Journal*, *313*(7059), 702–703.

Simmonds, D. J., Pekar, J. J., & Mostofsky, S. H. (2008). Meta-analysis of Go/No-go tasks, demonstrating that fMRI activation associated with response inhibition is task-dependent. *Neuropsychologia*, *46*(1), 224–232.

Simpson, E. H. (1951). The interpretation of interaction in contingency tables. *Journal of the Royal Statistical Society. Series B, Statistical Methodology, 13*(2), 238–241.

Simpson, R. J. S., & Pearson, K. (1904). Report on certain enteric fever inoculation statistics. *British Medical Journal, 1904*, 1243–1246.

Singh, I., & Rose, N. (2009). Biomarkers in psychiatry. *Nature, 460*(7252), 202–207.

Sirotin, Y. B., & Das, A. (2009). Anticipatory haemodynamic signals in sensory cortex not predicted by local neuronal activity. *Nature, 457*(7228), 475–479.

Smith, M. L., & Glass, G. V. (1977). Meta-analysis of psychotherapy outcome studies. *American Psychologist, 32*(9), 752–760.

Sommer, I. E. C., Aleman, A., Bouma, A., & Kahn, R. S. (2004). Do women really have more bilateral language representation than men? A meta-analysis of functional imaging studies. *Brain, 127*, 1845–1852.

Spreng, R. N., Mar, R. A., & Kim, A. S. N. (2008). The common neural basis of autobiographical memory, prospection, navigation, theory of mind, and the default mode: A quantitative meta-analysis. *Journal of Cognitive Neuroscience, 21*, 489–510.

Sternberg, S. (1969). Memory-scanning—mental processes revealed by reaction-time experiments. *American Scientist, 57*(4), 421.

Stroop, J. R. (1935). *Studies of interference in serial verbal reactions. Unpublished PhD thesis*. Nashville, TN: George Peabody College for Teachers.

Sudman, S., & Bradburn, N. M. (1973). Effects of time and memory factors on response in surveys. *Journal of the American Statistical Association, 68*(344), 805–815.

Svoboda, E., McKinnon, M. C., & Levine, B. (2006). The functional neuroanatomy of autobiographical memory: A meta-analysis. *Neuropsychologia, 44*(12), 2189–2208.

Talairach, J., & Tournoux, P. (1988). *Co-planar stereotaxic atlas of the human brain: 3-Dimensional proportional system—an approach to cerebral imaging*. New York: Thieme Medical Publishers.

Thompson, R. F. (2005). In search of memory traces. *Annual Review of Psychology, 56*, 1–23.

Thyreau, B., Schwartz, Y., Thirion, B., Frouin, V., Loth, E., Vollstadt-Klein, S., Paus, T., Artiges, E., Conrod, P. J., Schumann, G., Whelan, R., & Poline, J-B. (in press). Very large fMRI study using the IMAGEN database: Sensitivity-specificity and population effect modeling in relation to the underlying anatomy. *NeuroImage*.

Tian, T. S. (2010). Functional data analysis in brain imaging studies. *Frontiers in Psychology: Quantitative Psychology and Measurement, 1*, 1–11.

Tulving, E., Kapur, S., Craik, F. I., Moscovitch, M., & Houle, S. (1994). Hemispheric encoding/retrieval asymmetry in episodic memory: Positron emission tomography findings. *Proceedings of the National Academy of Sciences of the United States of America, 91*(6), 2016–2020.

Tunturi, A. R. (1952). A difference in the representation of auditory signals for the left and right ears in the isofrequency contours of the right ectosylvian auditory cortex of the dog. *Journal of Comparative and Physiological Psychology, 168*, 712–727.

Turkeltaub, P. E., Eden, G. F., Jones, K. M., & Zeffiro, T. A. (2002). Meta-analysis of the functional neuroanatomy of single-word reading: Method and validation. *NeuroImage, 16*(3), 765–780.

Tversky, A., & Kahneman, D. (1983). Extensional versus intuitive reasoning—the conjunction fallacy in probability judgment. *Psychological Review, 90*(4), 293–315.

Uttal, W. R. (1981). *A taxonomy of visual processes*. Hillsdale, NJ: Lawrence Erlbaum Associates.

Uttal, W. R. (1998). *Toward a new behaviorism: The case against perceptual reductionism*. Mahwah, NJ: Lawrence Erlbaum Associates.

Uttal, W. R. (2001). *The new phrenology*. Cambridge, MA: MIT Press.

Uttal, W. R. (2001). *The new phrenology:The limits of localizing cognitive processes in the brain*. Cambridge, MA: MIT Press.

Uttal, W. R. (2005). *Neural theories of mind: Why the mind-brain problem may never be solved*. Mahwah, NJ: Lawrence Erlbaum Associates.

Uttal, W. R. (2009a). *Distributed neural systems: Beyond the new phrenology*. Cornwall-on-Hudson, NY: Sloan.

Uttal, W. R. (2009b). *Neuroscience in the courtroom: What every lawyer should know about the mind and the brain*. Tucson, AZ: Lawyers and Judges Publishing Company.

Uttal, W. R. (2011). *Mind and brain: A critical appraisal of cognitive neuroscience*. Cambridge, MA: MIT Press.

Valera, E. M., Faraone, S. V., Murray, K. E., & Seidman, L. J. (2007). Meta-analysis of structural imaging findings in attention-deficit/hyperactivity disorder. *Biological Psychiatry, 61*(12), 1361–1369.

Videbech, P. (1997). MRI findings in patients with affective disorder: A meta-analysis. *Acta Psychiatrica Scandinavica, 96*(3), 157–168.

Villar, J., Carroli, G., & Belizan, J. M. (1995). Predictive ability of meta-analyses of randomised controlled trials. *Lancet, 345*(8952), 772–776.

Vul, E., Harris, C., Winkielman, P., & Pashler, H. (2009). Puzzlingly high correlations in fMRI studies of emotion, personality, and social cognition. *Perspectives on Psychological Science, 4*(3), 274–290.

Wager, T. D., Jonides, J., & Reading, S. (2004). Neuroimaging studies of shifting attention: A meta-analysis. *NeuroImage, 22*(4), 1679–1693.

Wager, T. D., Lindquist, M., & Kaplan, L. (2007). Meta-analysis of functional neuroimaging data: Current and future directions. *Social Cognitive and Affective Neuroscience, 2*(2), 150–158.

Wager, T. D., Lindquist, M. A., Nichols, T. E., Kober, H., & Van Snellenberg, J. X. (2009). Evaluating the consistency and specificity of neuroimaging data using meta-analysis. *NeuroImage, 45*(1), S210–S221.

Wager, T. D., Phan, K. L., Liberzon, I., & Taylor, S. F. (2003). Valence, gender, and lateralization of functional brain anatomy in emotion: A meta-analysis of findings from neuroimaging. *NeuroImage, 19*(3), 513–531.

Wager, T. D., & Smith, E. E. (2003). Neuroimaging studies of working memory: A meta-analysis. *Cognitive, Affective & Behavioral Neuroscience, 3*(4), 255–274.

Wernicke, C. (1874). *Der aphasische Symptomenkomplex.* Breslau, Germany: Cohn and Weigert.

Woolsey, C. N. (1952). *Pattern of localization in sensory and motor areas of the cerebral cortex: The biology of mental health and disease.* New York: Hoeber.

Yusuf, S., Collins, R., MacMahon, S., & Peto, R. (1988). Effect of intravenous nitrates on mortality in acute myocardial infarction: An overview of the randomised trials. *Lancet, 1*(8594), 1088–1092.

Zhang, J., & Yu, K. F. (1998). What's the relative risk? A method of correcting the odds ratio in cohort studies of common outcomes. *Journal of the American Medical Association, 280*(19), 1690–1691.

Name Index

Abbamonte, M, 119, 120
Abramoff, A., 122, 123, 125
Abramov, I., 43
Abrams, M. T., 93
Aguirre, G. K., 84, 86
Aleman, A., 37, 128, 161
Allen, P., 119, 120
Andersson, M., 196
Arbuckle, R., 94, 125
Armony, J. L, 128
Aron, A. R., 24
Artiges, E., 195
Asherson, P., 148
Astin, A. W., 13
Athwal, B. S., 85, 86, 201
Augath, M., 171

Baas, D., 128
Baker, C. I., 65
Bandetini, P. A., 195
Bangert-Drowns, R. L., 38
Bannon, K., 127
Baral, M.., 147
Barale, F., 36–37
Barrett, J., 122, 123, 125
Barrett, L. F., 104, 129, 130, 131, 150, 198, 202
Bearden, C. E.;, 122, 123, 125
Bellgowan, P. S. F., 65
Benhaddad, A., 76
Bennett, C. M., 24, 25, 26, 77

Berman, K. F., 161
Bessel, F., 12
Biddle, B. J., 13
Bliss-Moreau, E., 104, 129, 130, 131, 150, 198, 202
Bond, C. F., 14
Borenstein, M., 75, 76
Bornstein, R. F., 14
Bouma, A., 37, 161
Boyce, A., 93
Boynton, G. M., 204
Bradburn, N. M, 13
Bradstreet, J. J., 147
Brambilla, P., 36–37
Brammer, M. J., 160
Braun, B., 24
Brenowitz, N., 195
Broca, P., 150
Brodmann, K., 29
Brown, S., 38, 120, 121
Bruhns, P., 24
Buchsbaum, B. R., 161
Buckwalter, J. A., 146
Bullmore, E., 122, 124, 125, 126
Burdette, J. H., 24
Bush, G., 93, 161

Cabeza, R., 16, 30, 69, 96, 100, 101, 202
Cappell, K., 94
Carletti, F., 119, 120

Castellanos, F. X., 94, 125, 127
Caviness, V. S, 70
Chan, R. C. K., 119, 120
Chang, W. L., 161
Chein, J. M., 40, 43, 125
Chochol, C., 128
Chouinard, P. A., 160
Church, J. A., 65, 165, 202
Cohen, J., 197
Collins, D. L., 29
Collins, R., 74
Coltheart, M., 175, 176, 178
Conrod, P. J., 195
Constable, R. T., 31
Cooper, H. M., 14, 153
Corbetta, M., 169
Costafreda, S. G., 10, 160
Courchesne, E., 146
Cox, D. R., 135
Cox, R. W., 171
Craik, F. I., 15
Critchley, E. A., 19
Critchley, M., 19
Cunningham, W. A, 17, 72

Das, A, 59
David, A. S., 160
Deouell, L. Y., 43
Derderian, F, 76
Derrfuss, J., 49, 105, 106, 107
D'Esposito, M., 43, 84, 86
Devalois, R. L., 43
Devlin, J. T., 24
Dickersin, K., 63
Dickstein, S. G., 127
DiMatteo, M. R., 14, 53, 75
Dolan, R. J., 154
Donovan, C. L., 86, 88, 89
Driskill, L., 24
Duncan, K. J., 24
Dunkin, M. J., 13

Eden, G. F., 41, 43, 44, 45, 46, 94, 104, 111, 120, 121, 125, 135, 172

Eklund, A., 196
Ellermann, J. M, 166
Elliott, H. C, 30
Ellison-Wright, I., 126
Ellison-Wright, Z., 126
Engel, S. A., 204
Evans, A. C., 29
Everitt, B., 160
Eysenck, H. J, 12–13, 73, 76

Faraone, S. V., 126, 127
Feinstein, A. R., 73
Fesl, G., 24
Fiedler, K., 65
Fiez, J. A., 40, 43, 125
Fisher, R. A., 12
Fissell, K., 40, 43, 125
Fletcher, P., 154, 204
Flower, A. A., 93
Fox, M. D., 169
Fox, P. M., 106
Fox, P. T., 16, 120, 121, 153
Frackowiak, R. S. J., 71, 85, 86, 154, 201
Frazier, J. A., 161
Friston, K. J., 71, 85, 86, 154, 201, 204
Frith, C. D., 71
Frouin, V., 195
Fu, C. H. Y., 160
Fulbright, R. K., 31
Fusar-Poli, P., 119, 120

Gall, F. J., 2, 149
Garcia-Finana, M., 90
Gareau, L., 94
Gazzaley, A., 43
German, E., 86
Gilovich, T., 36
Glahn, D. C., 106, 122, 123, 125
Glass, G. V., 12, 13, 14, 15, 73, 197, 200
Glover, G. H., 204
Gluck, M. A., 24
Goldberg, M. C., 93

Gong, Q. Y., 119, 120
Gonzalez-Castillo, J., 195
Goodale, M. A, 160
Gould, S. J, 63
Greco, J. A., 71
Greer, S., 161
Gregoire, G., 76
Grunling, C., 94
Gusnard, D. A., 169
Guzzo, R. A., 14

Hajos, M., 146
Handwerker, D. A., 195
Handy, T. C., 86, 87, 88
Hanson, D. P, 16
Hardan, A., 36–37
Harris, C., 65, 66
Hartline, H. K., 200
Havel, P., 24
Hedges, L. V., 14, 75, 76, 153
Heeger, D. J, 204
Henson, R., 175, 176, 177
Higgins, J. P. T., 75, 76
Highley, J. R., 90
Hilgetag, C. C., 154
Hlatky, M. A., 73
Hogben, J., 68
Holmes, A., 204
Holmes, A. P., 71, 85, 86, 201
Holmes, J., 93
Houde, O., 160
Houle, S., 15
Howseman, A. M., 85, 86, 201
Hubel, D. H., 200
Hunter, J. E., 74, 75
Huonker, R., 94
Hutchinson, J. W., 70
Hyde, J. S, 171

Ihnen, S. K. Z., 65, 165, 202
Inati, S, 86, 87, 88
Inati, S. J., 195
Ingham, J. C., 120, 121
Ingham, R. J., 120, 121

Jacklin, C. N, 13
Jacobs, G. H., 43
Jacobs, S., 40, 43, 125
James, W., 67
Jasephson, C., 196
Javitt, D. C., 146
Jenike, M. A.;, 161
Jette, R. D., 14
Johannesson, M., 196
Johnson, M. H, 153
Joliot, M., 160
Jones, K. M., 41, 43, 44, 45, 46, 94, 104, 111, 120, 121, 125, 135, 172
Jonides, J., 47, 103
Joseph, J., 104, 129, 130, 131
Josephs, O., 204

Kahn, R. S., 24, 37, 128, 161
Kahneman, D., 36
Kao, M. H., 172
Kaplan, L., 10, 35, 47, 48
Kapur, S., 15
Karp, R. M.., 204
Katzell, R. A., 14
Keller, S. S., 90
Kemp, S., 2
Kennedy, D. P., 146
Killeen, P. R., 198
Kim, A. S. N., 117, 119
Kim, S. G., 166
Klingert, M., 94
Knierim, I., 24
Knutsson, H., 196
Kober, H., 9, 11, 34, 104, 129, 130, 131, 150, 198, 202
Kochunov, P., 105, 114, 115
Krain, A. L., 94, 125
Kriegeskorte, N., 65
Kringelbach, M. L., 69
Kuntsi, J., 148

Lacadie, C. M., 31
Laird, A. R., 105, 106, 114, 115, 120, 121, 122, 123, 125

Lancaster, J. L., 16, 105, 106, 114, 115
Land, E. H., 66
Landi, P., 119, 120
Lapierre, J., 76
Lashley, K. S., 156
Laurienti, P. J., 24
Lawrence, A. D., 31, 42, 129, 130, 131
Lazar, N., 172
Lee, D., 71
Lee, L., 160
Lehrer, J, 201
LeLorier, J., 76
Levine, B., 117, 118, 119
Levy, S. E., 147
Li, H. J., 119, 120
Liberzon, I., 16, 47, 203
Lieberman, M. D., 17, 72
Ligges, M., 94
Lindquist, K., 104, 129, 130, 131
Lindquist, K. A., 9, 11, 198, 202, 34150
Lindquist, M., 10, 35, 47, 48
Lipsey, M. W., 13, 200
Loftus, 198
Logothetis, N. K., 171
Lohmann, G., 49, 105, 106, 107
Loth, E., 195
Lubin, A., 160

Maccoby, E. E., 13
MacLeod, A. M., 169
MacMahon, S., 74
Maldjian, J. A., 24
Mandal, A., 172
Mandell, D. S.., 147
Mar, R. A., 117, 119
McAlonan, G. M., 119, 120
McGaw, B., 200
McGonigle, D. J., 85, 86, 201
McKinnon, M. C., 117, 118, 119
McMillan, K. M., 105, 114, 115, 122, 124, 125
Mehta, M. A., 148
Menon, R. S., 166

Mentzel, H. J., 94
Merkle, H., 166
Michell, J., 198
Milham, M. P., 94, 125, 127
Miller, M. B., 24, 25, 26, 77, 86, 88, 89
Milner, B., 29, 199
Modha, D. S., 11
Moore, C., 154
Moore, E. F., 180, 204, 206
Moscovitch, M., 15
Mostofsky, S. H., 92–93, 93
Mumford, J. A., 172
Munk, H., 150
Murphy, F. C., 31, 42, 129, 130, 131
Murray, K. E., 126, 127

Nee, D. E., 103
Neumann, J., 49, 105, 106, 107
Newman, S. D., 71
Nichols, T. E., 9, 11, 34
Nichols, T., 172
Nickerson, R. S., 198
Nimmo-Smith, I., 31, 42, 129, 130, 131
Nyberg, L., 16, 30, 69, 96, 100, 101, 202

Oeltermann, A, 171
Ogawa, S, 166
Okun, M. A., 38
Onarato, M., 169
O'Neill, M. A., 154
Osborne, A. R., 169
Owen, A. M., 122, 124, 125

Page, M. P, 179
Paloyelis, Y., 148
Papademetris, X., 31
Papez, J. W, 129, 152
Pardo, J. V., 105, 114, 115
Parsons, L. M., 16
Pashler, H., 65, 66
Pattamadilok, C., 24

Name Index

Pauls, J., 171
Paus, T., 16, 195
Pearson, K., 12
Pekar, J. J., 92–93
Perez, J., 36–37
Petersen, S. E., 65, 153, 165, 202
Peto, R., 74
Phan, K. L., 16, 47, 203
Picard, N., 16
Pierce, K., 146
Placentino, A., 119, 120
Poldrack, R. A., 24, 178, 191–192
Poline, J. P., 71
Poline, J-B., 195
Posner, M. I., 153
Powers, W. J., 169
Price, C. J., 106, 154, 161

Raemaekers, M., 24
Ragland, J. D., 122, 123, 125
Raichle, M. E., 153, 169
Rajeevan, N., 31
Ramsey, N. F., 24
Ratliff, F., 200
Rau, S., 24
Rauch, S. L., 161
Reading, S., 47
Redcay, E., 146
Rezaie, R., 90
Richman, D. P., 70
Robb, R. A., 16
Roberts, N., 90
Rolls, E. T., 69
Rose, N, 145
Rosen, B. R., 93
Rosenthal, R., 14, 53, 75
Ross, S., 13, 160
Rossignol, D. A., 147
Rothbart, M. K., 153
Rothstein, H. R, 75, 76
Rudner, L. M., 38
Rugg, M. D., 204
Rzanny, R, 94

Saad, Z. S., 195
Schafer, J. G. B., 93
Schlaggar, B. L., 65, 165, 202
Schmidt, F. L., 74, 75
Schramm, W., 13
Schultz, R. T., 147
Schumann, C. M., 146
Schumann, G., 195
Schwartz, Y., 195
Scoville, W. B., 199
Sears, D. O., 68
Seidman, L. J., 126, 127, 161
Sergerie, K., 128
Serio, M., 169
Shapiro, D., 13, 14
Shapiro, D. A., 13, 14
Shapiro, S, 73
Shercliffe, R. J., 14
Shin, L. M., 93
Shulman, G. L., 169
Sim, I., 73
Simmonds, D. J., 92–93
Simmons, W. K., 65
Simpson, E. H., 59, 74
Simpson, R. J. S., 12
Singh, I., 145
Singh, R., 11
Sirotin, Y. B., 59
Sluming, V., 90
Smith, E. E., 59, 125, 161
Smith, M., 200
Smith, M. L., 13, 73
Smith, S., 147
Smith, W. L, 135
Snyder, A. Z., 169
Soares, J. C., 36–37
Sokol-Hessner, P., 86
Sommer, I. E. C., 37, 161
Spencer, K. M., 146
Spreng, R. N., 117, 119
Spurzheim, J. C., 2, 149
Stahl, W., 14
Sternberg, S., 154

Stewart, R. M., 70
Strick, P. L., 16
Stroop, J. R.., 202
Stufken, J, 172
Sudman, S., 13
Svoboda, E., 117, 118, 119

Talairach, J., 29, 31, 70
Tank, D. W., 166
Taylor, S. F., 16, 47, 203
Thaker, G. K., 146
Thirion, B., 195
Thompson, R. F, 152
Thyreau, B., 195
Tian, T. S., 11
Titus, L. J., 14
Tonn, J. C, 24
Tournoux, P, 29, 31, 70
Trinath, T., 171
Tulving, E., 15
Tunturi, A. R., 150
Turkeltaub, P. E., 41, 43, 44, 45, 46, 104, 105, 111, 114, 115, 120, 121, 125, 135, 172
Turner, T, 204
Tuttle, M. P., 14
Tversky, A., 36

Ucelli di Nemi, S., 36–37
Uecker, A. M., 106
Uttal, W.R.., 6, 71, 102, 129, 155, 167, 173, 192, 197, 200, 202, 204, 205

Valentine, J. C.., 14
Valera, E. M., 126, 127
Vallone, R., 36
Valsangkar-Smyth, M.;, 86, 87, 88
Van Essen, D. C., 169
Van Horn, J. D., 86, 87, 88
Van Snellenberg, J. X., 9, 11, 34
van Wezel, R. J. A., 24
Videbech, P., 16
Vincent, J. L., 169

Vink, M., 24
Vogt, B. A., 93
Vollstadt-Klein, S., 195
von Cramon, D. Y., 49, 105, 106, 107
Vul, E., 65, 66

Wager, T., 16
Wager, T. D., 9, 10, 11, 34, 35, 47, 48, 59, 103, 104, 125, 129, 131, 139, 150, 161, 198, 202, 203
Wernicke, C., 150
Whalen, P. J., 161
Whelan, R., 195
Wiesel, T. N., 200
Wilson, A. M., 94, 125
Wilson, D., 13, 200
Winkielman, P., 65, 66
Winterer, G., 146
Woldorff, M. G.., 16
Wolford, G. L., 86, 87, 88
Wood, F. B., 94
Woolsey, C. N., 150
Worsley, K. J., 71

Young, M. P., 154
Yu, K. F., 38
Yusuf, S., 74

Zandbelt, B., 24
Zarahn, E., 84, 86
Zeffiro, T. A., 41, 43, 44, 45, 46, 94, 104, 111, 120, 121, 125, 135, 172
Zhang, J., 38

Subject Index

ACC (anterior cingulate cortex), 93, 107, 130, 161
Activation
  clustering of (*see* Clustering)
  concordant vs. irrelevant, 113
  of dorsal anterior cingulate cortex, 93
  in fMRI maps, 195
  frontal lobe, 92–93
  macroscopic patterns, 169
  meta-analyses comparisons, 117–126
  occipital, 88–89
  peaks (*see* Nodes)
  during retrieval conditions, 87–88
  sparseness, 109
  thresholds, 23, 69–70, 109
Activation likelihood estimate method (ALE method)
  as de facto method of choice, 111–112
  goal of, 45–46, 112–113
  regional clusters, 47
  Stroop test, 105–107
  subjectivity and, 46
  Venn fallacy and, 46
ADHD. *See* Attention deficit hyperactivity disorder (ADHD)
Aggregated Gaussian estimated sources (AGES), 40–41
ALE analysis. *See* Activation likelihood estimate method (ALE method)
Amygdala, 128–129

Anterior cingulate cortex (ACC), 93, 107, 130, 161
Artifacts
  case study, 199n1
  nodes as, 195
  from statistical manipulations, 27–28
  technical, 57, 68–69
  from two or more separate regions, 23
Attention, 83
Attention deficit hyperactivity disorder (ADHD)
  biomarkers, 146, 148, 161, 204n9
  narrative meta-analyses, 126–128
Autism biomarkers, 146–147, 204n9
Autobiographical memory, 117–119
Averaging
  clustering of average activations, 95
  of individual subjects (*see* Data pooling)
  pseudo-averaging, 21
  regional, 194
  sequence of, 113
  spatial, 21–22
  statistical, 18

BAs. *See* Brodmann areas
Basal ganglia, 127
Behavioral medicine meta-analyses, 14

Bias
  amelioration of, 51–52
  from decision criteria, 69–70
  in narrative interpretations, 36
  sources, 33, 54–57, 187, 193
  in voting, 37
Biomarkers, 140, 204n9, 205n10
  applications, 145
  biochemical, 147–148
  concept of, 144–145
  historical examples, 146
  neuroanatomical, 144–147, 148
BOLD (blood oxygen level
  dependence), 58–59, 166, 195
Brain
  activation (*see* Activation)
  anatomical interconnectedness of, 156
  anatomy, variability of, 70, 90
  cells (*see* Neurons)
  cognitively driven activity, 186–189, 192
  current metaphor for, 189
  environmental factors and, 204n4
  functional recovery of, 157–158
  individual differences in, 54
  locations, comparisons of, 29, 31
  macroscopic oscillatory activity, 169
  organization, 7, 11, 142
  response to cognitive stimuli, 50–51
  size, in ADHD, 126–128
Brain activation regions
  average/typical, 91
  broad/multiple, 51, 200n6
  Brodmann areas and, 156
  cognitive processes and, 4
  common, 21
  concordant, 113
  function-specific localization of (*see* Localization)
  non-overlapping, 21–23
Brain imaging-cognitive correlations
  as artifacts, 8
  reliability of, 192
  trust of, 111, 164–165
  variability and, 23–24
Brain imaging data, 2, 3
  as biomarkers (*see* Biomarkers)
  BOLD-determined, 165–170
  cognitively driven, 8
  comparisons, 33, 171–172
  correlations with cognitive processes
    (*see* Brain imaging-cognitive correlations)
  empirical, evaluation of, 83–84
  empirical reliability, 80
  increasing divergence of, 6–7
  interpretation of, 82, 201n2
  linking to cognitive processes, 190–191
  macroscopic, 191
  normalization, 171
  plotting of unpooled data, 104–109
  pooling of (*see* Data pooling)
  in psychology, 189
  in resolving theoretical controversies, 189
  spatial nature of, 142
  standardized nomenclature, 28–33
  theoretical psychology issues and, 172–179
  variability/inconsistency of, 1–3, 24, 139
Brain imaging techniques, 194–195. *See also specific techniques*
  disadvantages of, 5
  enhancement of, 170–172
  goal of brain mapping, 10
  limitations, 2
  limitations of instrumentation, 140
  macroscopic, 3 (*see also* Functional magnetic resonance imaging)
  nodes, methodological sensitivity of, 157
  research questions, 159–160

Subject Index

Brain maps
  Brodmann areas, 29, 30f, 156
  precision of, 70
  of raw activation regions, 188
  two-dimensional, 95, 101–103
Brain regions
  assignments, inconsistency in, 94
  coordinated/interconnected systems in, 4
  correlation's with cognitive activity, 188
  in declarative learning, 152–153
  in emotion, 152
  interconnected nodes, 23
  language, bilateral gender differences, 37
  macroscopic, 10
  macroscopic, relationship with cognitive processes, 140
  in memory, 152–153
  microscopic, 10
  modular psychological functions and, 3–4
  multifunctionality of, 156
  multiple functions for, 187
  node concept, 156 (see also Nodes)
  size in dysfunctional behavior, 204n6
  specialized/circumscribed, 2–3
  in visual attention, 153
Brain surgery, 149–150, 201n5
Broca's area, 19–20, 40, 90
Brodmann areas (BAs), 31, 33, 95, 121, 131
  activation peaks, 103, 116
  brain map, 29, 30f, 156
  executive function and, 161
  in n-back working memory, 122–124
  in working memory, 124–126

Case studies (single-subject experiments), 33, 84–86, 142
Caudate, 127
Central tendency, 2, 20

Chladni plate, 169
Cingulate gyrus
  activation peak distribution patterns, 95, 96–101
  in ADHD, 127
Closed systems, internal mechanisms of, 206n20
Clustering, 110, 132, 199–200n4
  of activation peaks, 42, 112
  of average activations, 95
  common impact of stimulus and, 39–40
  imperceptible, 201n6
  regional, 44–45
Coding schemes, 62
Coefficients of variation, 24
Cognitive-brain associations, 35
Cognitively generated activity, broad distributions of, 188
Cognitive neuroscience
  data (see Brain imaging data)
  emerging principles, 186–190
  future needs, 190–195
  issues, 185
  meta-analyses, history of, 12
  research, 5–6
  theories, 141, 159–163, 189 (see also specific theories)
Cognitive process-brain image correlations, 187
Cognitive processes, 70, 83, 92
  activation, variability in, 19
  associations with brain regions, 187
  autobiographical memory, 117–119
  brain encoding of, 7–8
  broad brain systems for, 190
  correlation with brain images (see Brain imaging-cognitive correlations)
  definitions, variability of, 54
  emotional face perception, 119–120
  higher, 198n8
  high-level, 187
  holding constant, 67

Cognitive processes (cont.)
  inferring from specific brain regional activations, 192
  interconnectivity of, 4
  logical chain from BOLD to cognition, 165–168
  n-back working memory, 122–124
  neural coordinates, 200n10
  single-word reading, 120–122
  taxonomy, 192
Comparability, 116
Computer vision processing, 40–41
Conceptualization errors, 56, 58–60
Concordance, 203n15
Confounding variables, 67
Consistency, 113–114, 201n3
Contrast effects, 66
Contrast level of image, 69
Control conditions, 62, 65, 188
Convergence, 94
Cortex (journal), 175
Cortical regions, emotional behavior and, 129, 130
Cutoff frequency, 71

Data pooling methods, 8, 9, 27, 35, 52, 113
  basic assumptions of, 6
  convergence of results, 9, 80–82
  dimension reduction and, 11
  distortions in, 71–72
  equivalence of experiments for, 54
  historical background, 12
  inappropriate, 60
  interstudy differences and (*see* Interstudy differences)
  levels, 188
  localization theories and, 3
  mega-analyses, 198n7
  metrics, 20–21
  need for, 9, 33
  objective/subjective decision making for, 61–62
  problems with, 19–26
  purpose of, 5
  quantification of, 41–42
  results, criticism of, 73–77
  sequence, 113
  similarities of data for, 54
  simple comparisons, 94–95
  statistical power and, 5, 19, 187–188
  successive, 55
  validity of, 132
  variability, 6
Decision criteria, 57, 69–70
Decision making tasks, 94
Descriptive statistics, 38
Diffusion tensor-type imaging, 11, 156
Dimension reduction, meta-analyses and, 11
Distribution, 7, 159, 169, 200n6
  alternative version, 157–158
  evidence, 151, 205n11
  with function-specific nodes, 151–155
  glass brain, 104–109
  tabular presentations, 95, 96–101
  two-dimensional plots, 101–103
  vs. localization, 10–11, 149
  without function-specific nodes, 155–158
Double dipping error, 65–66
Duplicates, 62

EEG, 186, 200n9
Effect size, 38, 39
Emotional face perception, 119–120, 128
Emotional stimuli, 31, 32, 198–199n13
Emotions
  brain locations for, 198–199n13
  brain regions of, 129–130
  neural mechanisms of, 152
  use of term, 83
Episodic encoding/retrieval, 89
Episodic memory, 89
Episodic memory retrieval task, 86–88

Subject Index

ERP, 200n9
Errors
  case study, 33
  of conceptualization, 56, 58–60
  of extrapolation, 59
  interpretive, 22
  in original experimental design, 56, 64–66
  sources, 54–57, 187
  statistical/methodological, 71–72
  type I, 17, 18, 28, 59
  type II, 17, 18, 19, 28
  Venn fallacy, 22, 23
Event-related fMRI, 18
Event-related potential (ERP), 18
Executive decision making, Go/No-go test, 92–93
Experimental conditions, 193
Experimental design, 110
  confounded factors, 64
  to control type I errors, 72
  original, errors in, 56, 64–66
Experiment selection process, 113
Extrapolation errors, 59

False positives, 35
"File drawer problem," 74
fMRI. *See* Functional magnetic resonance imaging (fMRI)
Frontal cortex, 160
Frontal lobe
  activation, 85, 92–93
  activation peaks, 95–101, 202n11
  ADHD and, 127
  glass brain and, 107
Functional magnetic resonance imaging (fMRI), 189, 191
  of activation peak distribution patterns, 95, 96–101
  artifacts, technical, 69
  bias, 108
  blood oxygen levels measurement, 166–168

  of cognitive tasks, 150
  correlations with cognitive reactions in social situations, 65
  in determining biomarkers, 147
  development, 186
  differences in decision making tasks, 94
  episodic memory retrieval task, 86–88
  gender differences, 65
  localization, variability in, 93
  maps, activations in, 195
  meta-analyses, 16–17, 39
  neuronal networks and, 59
  reliability, 24
  sensitivity, 70–71
  signal analysis, 71
  for single-word-reading task, 45
  temporal dynamics, 161, 204n5
  test-retest reliability, 24
Function specificity, 162

"Garbage in, garbage out" criticism, 76
Gaussian filters, 71
Gaussian probability distribution, 48
Gaussian space, 112
Gaussian weighting function, 199n3
Gender differences, 65
Glass brain depiction, of activation peaks
  description of, 95, 104–109
  for single-word reading task, 43, 44
Globus pallidus, 127
Go/No-go test, 92–93, 161
Gray literature, 63–64
Group averages, 187

Hazard ratios, 38
Hebbian concept of the mind, 194
Hebb model, 186
Hippocampus damage, 199n1
Human-mechanical system interaction, 54
Hypothalamus, 152

ICC (interclass correlation coefficient), 116
Illusions, 66
Inconsistency, 82, 187. *See also* Variability
Individual differences, 22, 33, 84–91
Individual experiments, 110
Interclass correlation coefficient (ICC), 116
Interconnectivity of brain, 11
Interday replications, 67
Interexperiment variability, 187
Interpretation, of pooled data, 82
Interstudy differences, 91–109
Intersubject reliability, 92
Intersubject variability, 20, 187, 188
Intraclass correlation coefficient (ICC), 24, 25
Intrasubject variability, 187
Inverse chi-squared method, 12

James-Lange theory, 205n12

Kernal density analysis (KDA), 47–49

Language
  functional lateralization, 161
  lateralization between men and women, 199n2
Lateral orbitofrontal region, 130
Learning
  declarative, brain regions of, 152–153
  use of term, 83
Left inferior frontal gyrus, 160
Levels of analysis, 142, 194
Localization, 34, 162, 190, 202n9
  as dominant postulate, 159
  fMRI, variability in, 93
  lack of evidence for, 3, 5, 150, 154–156
  as misleading, 7–8
  narrow, 19–20, 23
  neophrenological, 150
  phrenological, 51
  probability distributions, 43
  pseudo-averaging, 21
  regional, 4
  in sensory and motor domains, 150
  specificity of, 9–10
  theories, 3, 149–150
  of thought processes, 153
  three-dimensional (*see* Glass brain)
  vs. constructivist approach, 198n10
  vs. distribution, 10–11
Localized nodes, 148–151

Macroscopic theories of mind-brain relations
  connectionist, 195
  historical aspects, 149–150
  methodology, 170–183
  types, 143–158
Magnetic resonance imaging machine (MRI), 68–69
Mantel-Haenszel statistics test, 16
Mathematical models, 42, 181–182, 206n18
Maximum likelihood estimates, 24
Measurement levels, 58
Mega-analyses, 198n7
Memory
  autobiographical, 117–119
  brain regions in, 152–153
  failure, short-term, 199n1
  working (*See* Working memory)
Meta-analyses, of brain image data, 15–16. *See specific aspects of*
  advantages/disadvantages, 5–6, 73–77, 197n5
  assumptions, key, 11
  bias sources, 50–57, 193
  classic or Glassian, 38
  comparing for consistency (*see* Meta-meta-analysis)
  conceptual foundation of, 111
  conclusion from, 6

Subject Index

criticisms of, 14, 73–77
definition of, 187–188
determination of brain region interconnectivity, 11
differences between, 188
dimension reduction and, 11
disadvantages, 6
error sources, 50–57
findings from, 131
of fMRI data, 16–17
goal of, 112–113
goals of, 9–10, 42
historical overview, 12–17
limitations/problems, 58
methodology, 27, 35–50, 49–50 (*see also* Data pooling; *specific types of meta-analysis*)
spatial patterns, 39–41
typology of, 33–35
need for, 27
of neuroscientific findings, 15
PET scans, 15–17
premise, basic, 9
psychological, 12, 13, 14–15
purpose of, 9–11, 18
"recipes," steps in, 53
standardization, 14
typical, 27, 110–111
use of term, 12, 197n6
validity of, 59
variability in, 50–57
vs. mega-analyses, 198n7
vs. single-studies, 75–76
Meta-meta-analyses, 114
autobiographical memory, 117–119
design issues, 109–117
emotional face perception, 119–120
narrative, 126–132
n-back working memory, 122–124
single-word reading, 120–122
working memory, 124–126
Methodological errors, in meta-analysis, 57, 71–72

Mind-brain relationship, 52, 58, 194–195
brain imaging studies, 191
distributed brain function (*see* Distribution)
group averages and, 90
localized brain function (*see* Localization)
macroscopic theories, 139–183, 140–141
microscopic theory, 140–141
models, 200n9
neuronal connections, 202–203n14
random complexity of, 141
theories, 204n8
MKDA (multilevel kernal density analysis), 47, 48–49
MNI (Montreal Neurological Institute), 31, 43, 71
Moderator variables, 74
Modularization, 5, 10
Monte Carlo calculations, 48
Montreal Neurological Institute (MNI), 31, 43, 71
Motor systems, 198n8, 200n10
MRI (magnetic resonance imaging machine), 68–69
Multilevel kernal density analysis (MKDA), 47, 48–49
Multi-Source Interference Test, 93

Narrative descriptions, of spatial information, 115–116
Narrative interpretations, 35–36
n-back working memory, 122–124
Neural correlates, of cognitive activity, 1
Neurobiological variability, 70–71
Neurons, 140, 186
brain activity data, pooling of, 165–168
interconnections, 142, 202–203n14
Neurophysiological measures, 189

Node-based theories, 196
Nodes (activation peaks), 42, 85, 110, 195
  as artifacts, 195
  distribution patterns, 95–96, 104–109
  emotional stimuli, 31, 32
  fMRI meta-analyses, 39
  function-specific, 162–164
    distribution with, 151–155
    lack of evidence for, 154–155
  Gaussian weighted zones around, 40
  weakly bound, 156
Noise, 19, 112, 189, 193
  neurobiological, 23
  vs. signal, 10, 17–18, 47, 165
Non-Brodmann areas, 119, 123
Null hypothesis rejection, 193

Object tasks, activation peak distribution patterns, 98–99
Occipital lobe
  activation peaks, 95, 96–101, 202n11
  ADHD and, 127
Occipital visual area, 150
Odds ratio, 38
Orbitofrontal region, 130
Outcome of meta-analysis, 76, 111, 113

Parietal cortex, 160
Parietal lobe
  activation peak distribution patterns, 95–101
  ADHD and, 127
  left inferior lobule, 107
Pars opercularis, 90
Pars triangularis, 90
Permutation-type experiments, 202n13
PET scans
  of activation peak distribution patterns, 95, 96–101
  of cognitive tasks, 150
  meta-analyses, 15–17
  nontherapeutic, 34
  technical artifacts, 69
Phonological processes, 160
Phrenology, 149
p levels, 193
Pooling of data. See Data pooling
Power of analyses. See Statistical power
Presupplementary motor area, 107
Probability, 67, 69
Probability distribution regions, 43
Progressive pooling, 117
Prototheory, 188
Psychoanalytic therapy, 13
Psychobiology, 92, 189, 193
Psychological theories, 181, 189, 206n21
Psychology, 189
  dimensions, 20
  experimental, 37
  history of meta-analyses in, 12–15
  modular functions, 3–4
  research, statistical significance in, 25
  theoretical questions, resolution of, 172–179
  vocabularies, definitional softness and, 131
Psychometric meta-analyses, 38
Psychotherapy, 12–13
Publication bias, 62–64
Pure insertion, 154, 174
Putamen, 127

Quantitative theory, 188
Quasi-randomness, 203n2
Questions, 198n9

Reading speed, 202n12
Relative risk, 38
Reliability, 55, 79, 80, 193
  of brain images, 24–26
  definition of, 22
  interstudy differences (see Interstudy differences)

Subject Index

of narrative meta-meta-analyses, 126
numeric score, 116
of single-subject experiments, 84–86
Replicability, 193
Replicator dynamics, 49
Response patterns, 110, 111, 187
Reverse inference, 192
Robustness, 194
Rostral supracallosal anterior cingulate, 130

Samples, unrepresentative/restricted, 68
Sample size
 inadequate, 27
 larger/increased, 6, 8, 51
 signal-to-noise ratio and, 18
 significance and, 18
 small, 8, 34, 37
 statistical power and, 68, 197n4
Schizophrenia biomarkers, 146
Search process, inadequate, 60–61
Second-stage filter, 92
Selection biases, 56, 60–64
Sensory studies, 66–67
Sensory systems, 198n8, 200n10
Sex differences, 161
Short-term memory failure, 199n1
Signal detection theory, 42
Signals
 average, 55
 definition of, 112
 vs. noise, 10, 17–18, 47, 112
Signal-to-noise ratios, 43, 171
 irregular, 109
 low, 52, 91, 111, 170
 sample size and, 18
Significance
 of pooling insignificant data, 72
 sample size and, 18
 testing, 14–15
Significant differences, 27

Simpson paradox, 59, 74, 76, 132
Single-neuron studies, 186
Single-subject experiments (case studies), 33, 84–86, 142
Single-word reading
 meta-analysis comparisons, 120–122, 202n11
 PET studies, 104
 statistically significant brain regions, 44
Social neuroscience, 65
Social sciences, 12
Somatosensory homunculus, 150
Spatial arrangement, in sensory study, 66
Spatial averaging, 21–22
Spatial information presentation, 115–116
Spatial patterns, 39–41
 localization of (*see* Localization)
Spatial task activation peak distribution patterns, 99–100
Speech areas, 150
Speech production, 161
Split-half designs, 115
SPM2 (statistical parametric mapping), 71
Statistical analysis, 9
 artifacts from, 27–28
 averaging of pooled data, 37–38
 conventional, 37–38
 errors, 57, 71–72
 errors, types of, 17, 18
 limitations of, 54–55
 methods of, 191, 206n18
 significance testing, 14–15
Statistical distributions, 20
Statistical parametric mapping (SPM2), 71
Statistical power, 5, 11, 27, 34, 187, 192
 data pooling and, 5–6, 19
 definition of, 197n4

Statistical significance, 193
Stimulus (stimuli), 52, 66–67
Stroop test, 105–107
Study effect meta-analyses, 38
Subcortical regions
  activation peak distribution patterns, 95, 96–101
  emotional behavior and, 129–130
Subjectivity, 46
Subjects
  behavioral relationship with MRI machine, 68–69
  individual variability of, 54
  internal influences, 67
  variability of, 66–68

Tabular presentations, of activation peak distribution patterns, 95, 96–101
Talairach-Tournoux coordinate system, 29, 31–33, 70, 95, 112, 116
Temporal lobe, 95, 96–101, 202n8
Tests of homogeneity, 38
Thalamus, 127
Theories, 205n14, 206–207n22
  of brain organization, 142
  cognitive neuroscience, 189
  descriptive, 182
  distributed (see Distribution)
  explanatory, 182
  flawed explanations from, 33
  macroscopic connectionist, 195
  necessity/sufficiency of, 179–183
  a priori beliefs or zeitgeist, 28, 58, 109, 196
  psychological, 181, 189, 206n21
  psychological, neurophysiological resolution of, 172–179
Thresholding process, 40
Thresholds, 23, 69–70, 109
Two-dimensional plots, of activation peak distributions, 101–103

Validity, 111
Variability, 35, 37, 110, 193
  anatomical, 90
  of brain imaging, 5, 9, 25
  cognitive, 56
  experimental conditions and, 7, 11
  of individual brain response, 23
  intersubject, 33, 84, 86–87, 90–92
  intrasubject, 23, 84, 86–87, 90–92
  major source of, 66
  neurobiological, 25, 27, 57, 70–71
  sources, 39
  of statistical distributions, 20–21
  voting procedure and, 37
Venn fallacy, 22, 23f, 46
Venn logical approach, 22–23
Verbal/numeric tasks, activation peak distribution patterns, 96–97
Visual attention, brain regions in, 153
Voting procedures, 13–14, 36–37, 199n2

Wisconsin card sorting task (WCST), 161
Working memory
  activation peak distribution patterns, 95, 96, 101
  brain activation meta-analyses comparisons, 124–126
  meta-meta-analyses, 124–126
  verbal, left frontal dominance of, 161

Zeitgeist, 28, 58, 109, 196